PLUMBING

NVQ DIPLOMA LEVEL 2

Eamon Wilson

Nelson Thornes

Published in 2011 by:
Nelson Thornes Ltd
Delta Place
27 Bath Road
CHELTENHAM
GL53 7TH
United Kingdom

11 12 13 14 15 / 10 9 8 7 6 5 4 3 2 1

A catalogue record for this book is available from the British Library

ISBN 978 1 4085 0882 4

Cover photographs: David White/iStockphoto (gold-plated taps); Diego Cervo/iStockphoto (construction worker); Pali Rao/iStockphoto (hand-spanner)
Illustrations by Pantek Media
Page make-up by Pantek Media, Maidstone

Printed and bound in Spain by GraphyCems

Acknowledgements
B&Q 92lm; **Brett Martin Ltd** 180t, 180tm, 180lm, 180b; **Coram Showers Ltd** 191; **Eamon Wilson** 33(5), 190; **Flameport Enterprises** 92tm; **Fotolia.com** (Aleksandar Jovanovic) 156(1), (M1DSU) 107t, (Roman Kholodov) 107b; **Getty Images/ Image Source** 1; **Grundfos Pumps Ltd** 138b; **Heatrae Sadia** 73t; **Honeywell Controls Systems Ltd** 49(4), 50(7), 138m, 155(4), 155(5), 156(2), 156(3), 156(5), 156(7); **iStockphoto** (Adam Borkowski) 9E, (ajaykampani) 73b, (Andrea Pelletier) 33(2), (Anthony Seebaran) 9C, (Ary6) 192, (asterix0597) 33(3), (Benjamin Brandt) 81, (Bradley Mason) 49(1), (Branko Miokovic) 42, (Branko Miokovic) 41, (Dan Wilton) 33(1), (Darya Petrenko) 49(3), (Dmitry Naumov) 155(2), (f_) 9B, (Gordon Saunders) 26, (Günay Mutlu) 9D, (Ivan Chuyev) 155(3), (Joe Gough) 156(4), (Joe Michl) 130, (Jonathan Maddock) 22(3), (Kamo) 171, (Karl-Friedrich Hohl) 33(4), (kelvinjay) 50(5), (Mark Richardson) 22(1), (Michael Sleigh) 49(2), (MightyIsland) 162, (Nicholas Belton) 193l, (nsj-images) 100, (nsj-images) 155(1), (Owen Price) 103, (P_Wei) 9A, (Pali Rao) 92t, (Pedro Castellano) 47, (Pedro Castellano) 60, (PLAINVIEW) 116, (sturti) 145, (Tammy Bryngelson) 22(2), (tirc83) 193r; **Kamco Ltd** 50(6); **Polypipe Building Products** 181; **Reliance Water Controls Ltd** 138t; **Ridge Tool Company** 113; **Salt London/Wickes** 23; **Schneider Electric** 92b; **TFC Group – Grasslin (UK)** 156(6); **Toolbank Marketing Services/Dormole Ltd** 108t, 108b

Contents

Introduction

Welcome to the *Plumbing Level 2 Course Companion*. It is literally a companion to support you throughout your course and record your progress!

This workbook style book is designed to be used alongside any student book you are using. It is packed full of activities for you to complete in order to check your knowledge and reinforce the essential skills you need for this qualification.

Features of the *Course Companion* are:

Unit opener – this page contains a brief introduction to each unit along with the learning objectives you need to achieve

Key knowledge – the underpinning knowledge you must know is summarised

Activities – a wide variety of learning activities are provided for you to complete in your *Companion*. Each activity is linked to one of the Personal, Learning and Thinking Skills to help you practise these fundamental skills:

– Reflective Learner

– Self Manager

– Creative Thinker

– Independent Enquirer

– Teamworker

– Effective Participator

You will also notice additional icons that appear on different activities, which link to the following core skills and also to rights and responsibilities in the workplace:

– Literacy

– Numeracy

– ICT

– Employment, Rights and Responsibilities

Key terms – during your course you'll come across new words or new terms that you may not have heard before, so definitions for these have been provided

Your questions answered - your expert author, Eamon Wilson, answers some burning questions you may have as you work through the units

Quick quiz – At the end of each unit you will find a quiz. Answering these questions will check that you have fully understood what you have learnt.

Good luck!

Unit 001/201

Understand and carry out safe working practices in building services engineering (BSE)

Health and safety is an essential part of everybody's working life. It could mean the difference between life and death.

The Health and Safety Executive (HSE) compiled the following statistics:

- Construction has the largest number of fatal injuries of the main industry groups. In 2009/10, there were 42 work-related deaths (2.2 per 100,000 workers). This is the third highest rate of fatal injuries, after agriculture and the extractive industries (mining and quarrying).
- Construction accounted for 35 per cent (276 cases) of all reported injuries involving high falls and 24.8 per cent (89 cases) of cases involving electricity.
- In 2009/10, 3.3 million working days were lost in construction due to workplace injury and work-related ill health.

Source: www.hse.gov.uk/statistics/industry/construction/index.htm

It is the responsibility of both the employer and the employee to reduce the risk of accidents by adhering to health and safety legislation. For instance, the employer should provide safe working conditions and ensure all staff are trained in health and safety procedures. The employee has to follow health and safety procedures and ensure appropriate personal protective equipment (PPE) is worn on site.

Learning outcomes

- Know the health and safety legislation that applies to the building services industry
- Know how to recognise and respond to hazardous situations while working in the building services industry
- Know the safe personal protection measures while working in the building services industry
- Be able to apply manual handling techniques
- Know how to respond to accidents that occur while working in the building services industry
- Know the procedures for electrical safety when working in the building services industry
- Be able to apply basic electrical safety measures in the building services industry
- Know the methods of working safely with heat producing equipment in the building services industry
- Be able to safely work with gas heating equipment in the building services industry
- Know the methods of safely using access equipment in the building services industry
- Be able to safely use access equipment in the building services industry
- Know the methods of working safely in excavations and confined spaces in the building services industry

Key knowledge

- Health and safety laws and regulations
- Health and safety terms
- How to keep a working environment safe for everyone
- Understand what is meant by hazards and the regulations behind them
- Understand different types of fire extinguishers
- Accident procedures and prevention
- Basic electrical safety on site
- Working safely at height and in excavations

key terms

Regulations: compulsory rules set out to control and manage procedures within any industry.

Health and safety legislation in the building services industry

ACTIVITY

Health and safety legislation

There are 11 main health and safety **regulations** that affect the construction industry. The titles are listed below with a brief description; your task is to match up the appropriate description to the regulation by putting in the correct number that you think matches the letter. For example, if you think box A combines with description number 2, then your answer in the box would be A2. We have completed one as an example.

Regulation	Description	1st attempt	2nd attempt	Tutor to tick if correct
Health and Safety at Work Act **(A)**	**(1)** Main rules that cover health and safety in the workplace To provide safety in the workplace Protects visitors and the public	A ______		
Provision and Use of Work Equipment Regulations **(B)**	**(2)** Rules that cover dangerous solids, liquids or gases and how they should be used and stored Actions you and your employer must take to protect your health and the health of others	B ______		
Personal Protective Equipment at Work Regulations **(C)**	**(3)** A legal framework that covers all aspects of electrical work and equipment to be used You must follow these rules – failure to do so could lead to prosecution	C ______		
Control of Substances Hazardous to Health (COSHH) Regulations **(D)**	**(4)** Includes hard hats, impact goggles, ear defenders, protective gloves, overalls, safety footwear and high visibility clothing	D ______		

Regulation	Description	1st attempt	2nd attempt	Tutor to tick if correct
Workplace (Health and Safety at Work Regulations) **(E)**	**(5)** These regulations require employers to: • analyse workstations to assess and reduce risks • ensure workstations meet specified minimum requirements • plan work activities so that they include breaks or changes of activity	**E** ______		
Management of Health and Safety at Work Regulations **(F)**	**(6)** This is a legal requirement to do with your health and safety within the workplace Your employer must carry out a risk assessment Your employer must act upon risks to minimise them	**F** ______		
Manual Handling Operations Regulations **(G)**	**(7)** These regulations aim to ensure that workplaces meet the health, safety and welfare needs of all members of a workforce, including people with disabilities	**G** ______		
Electricity at Work Regulations **(H)**	**(8)** The regulations require mandatory training for anyone liable to be exposed to asbestos fibres at work (see regulation **J**). This includes maintenance workers and others who may come into contact with or who may disturb asbestos (e.g. cable installers), as well as those involved in asbestos removal work	**H** ______		
Working at Height Regulations **(I)**	**(9)** These regulations place duties upon employers in respect of their own employees. Identical duties are placed on self-employed workers in respect of their own safety These regulations seek to reduce the very large incidence of injury and ill-health arising from the manual handling of loads at work. More than one in four of all reportable injuries are caused by manual handling	**I** ______		
Control of Asbestos at Work Regulations **(J)**	**(10)** This document summarises what you need to do to comply with the Work at Height Regulations 2005. Some industry/trade associations may have produced more detailed guidance about working at height	**J** 8		
Display Screen Equipment at Work Regulations **(K)**	**(11)** Provides guidance to protect people's health and safety from equipment that they use at work Equipment that includes electric drilling machines and any piece of equipment provided by you or your employer	**K** ______		

ACTIVITY

In the following situations, which of the regulations in the previous activity have been broken?

1. A plumber fails to wear a hard hat on site.

2. Mains voltage equipment is used on a construction site.

3. Ceiling tiles containing asbestos are thrown into the skip.

4. An employee is working on a ladder with a broken rung.

Identifying and dealing with hazards in the work environment

Employers are required to provide safety signs in a variety of situations. There are four different types of safety signs:

- prohibition
- warning
- mandatory
- safe way to go.

In addition, there are symbols that are used on the packaging of hazardous substances to comply with the Chemicals (Hazard Information and Packaging for Supply) Regulations 2009. These are black pictures on an orange background (COSHH).

Recognising health and safety symbols

In this task you should categorise the symbol, give its meaning and an example of a situation it may cover. The first one has been done for you.

You can use the internet to help you.

Symbol	Type	Meaning	Situation
	Warning sign	*Explosive*	*Leaking gas main*
OXIDISING			
TOXIC			
IRRITANT			
DANGEROUS FOR THE ENVIRONMENT			

key terms

Hazard: a situation that may be dangerous and has the potential to cause harm.

REFLECT

ACTIVITY

Describe THREE common situations that could constitute a **hazard** in the workplace and what should be done to minimise risk.

1.

2.

3.

For example, Jenni is working at height fixing a downpipe. She has put up a warning sign telling others that work above is taking place and she has also put up a barrier with mandatory signs explaining to anyone who enters the area that they must wear the correct PPE.

COSHH

COSHH stands for the Control of Substances Hazardous to Health. These regulations cover the use of chemicals and other hazardous substances at work that can put people's health at risk. You might not think that this would apply within the plumbing industry, but if you think about the procedure you follow for dealing with waste pipe work, for instance, you use solvents, which would come under these regulations.

ASK

ACTIVITY

For this task you will need to go into the workshops and locate chemicals you use in plumbing.

Answer the questions in the table below for each of the chemicals. You can find the information by reading the labels, asking questions and looking up the substances in books or on the internet.

Substance name	What is the chemical used for?	How can this chemical cause harm?	How can you reduce the risk? (Time using product)

Substance name	What is the chemical used for?	How can this chemical cause harm?	How can you reduce the risk? (Time using product)

Reporting of Injuries, Diseases and Dangerous Occurrences Regulations 1995 (RIDDOR95)

RIDDOR requires the reporting of work-related accidents, diseases and dangerous occurrences. The regulations apply to all work activities, but not to every incident. The reporting of these serious accidents and any instances of ill-health due to work activities is a legal requirement. This information enables the authorities to identify and examine how these risks and accidents arise, and then if they need to implement new **legislation** to help prevent the recurrence of these problems. Accidents that mean you are off work for three days or more, a dangerous occurrence that may not have resulted in an accident but could have done and disease that has been verified by a doctor must all be reported.

key terms

Legislation: the process for making new laws and the laws themselves.

ACTIVITY

List as many relevant reportable diseases/accidents or occurrences that you can think of that you may come across in your work as a plumber.

Asbestos

Asbestos is a naturally occurring fibre. It is processed in numerous ways for industrial use. It is mainly used within the plumbing industry for its properties of **resistance** to heat.

There are THREE types of asbestos: what are they and what colour are they?

1.

2.

3.

Unfortunately, asbestos is also dangerous to health. Breathing in asbestos fibres can lead to the development of THREE fatal diseases: what are they? Use the internet to obtain the information:

1.

2.

3.

key terms

Resistance: deflecting the heat away and not transferring heat to a material behind or around it.

ACTIVITY

Asbestos – where is it?

In groups, discuss the various places where you as a plumber could find asbestos. Write up your findings.

Personal protective equipment (PPE)

Personal protective equipment (PPE) is essential for every plumber and mechanical engineer and it is important that you know which PPE you should use and when.

key terms

Personal protective equipment (PPE): all the equipment you use to keep yourself safe from having an accident.

ACTIVITY

Name the items of PPE pictured and describe what each is used for.

A	B	C	D	E
Name: ______	Name: ______	Name: ______	Name: ______	Name: ______
Used for: ______	Used for: ______	Used for: ______	Used for: ______	Used for: ______

REFLECT

ACTIVITY

Give THREE reasons why PPE is important and describe what might happen if PPE isn't used or is not used properly.

1.

2.

3.

What might happen if PPE is not used/not used properly?

ACTIVITY

In small groups, discuss the following tasks that are to be carried out on site by various workers and decide which items of PPE should be used.

Task	PPE needed to carry out task
Mohamed has been asked to guide the merchant's truck on to the site	
Tim is on a building site, helping to unload barrel pipe	
Emmy is using a circular saw to cut the notches for the pipework	
Alanna is using her 110v SDS drill to make fixings for clips	

Manual handling

You might have already demonstrated as a plumber that you can lift heavy objects such as a bath or a boiler, but manual handling regulations will even cover your tool box. Take a couple of minutes to weigh your tool box on your scales at home – I think you will be quite surprised how heavy it is.

You may also be surprised that the regulations do not give any specific weight limits. What they do cover is something called an '**ergonomic approach**'. Having a specific limit of weight for each gender could cause problems and may possibly lead to inaccurate conclusions. Instead, you and your employer must determine the risk of injury for lifting or moving something heavy.

key terms

Ergonomic approach: taking an educated approach to a situation; the word actually translates to mean 'science of work'.

ACTIVITY

Describe what you would do in these THREE situations and state if the work is safe to complete.

1. You have been asked to move a box with a warning sign showing that it weighs 30 kg.

2. You are instructed to lift a boiler on to its bracket but you don't know how much it weighs.

3. Your workmate needs help with moving a bath upstairs.

Health and safety procedures and how to deal with accidents

In order to comply with health and safety legislation, all construction sites have to undertake standard procedures and inform site **operatives** of these relevant procedures for the site. Visitors and **contractors** always have to be given a site induction by the site manager or foreman. You will also have to carry a **CSCS** card to be let on to site. This proves you have passed a health and safety test and that you are deemed to be aware of health and safety on site.

key terms

Operatives: all persons who are working on site.

Contractors: persons who are on site but go to and from different sites as they are not 'employed' by a company but work to a contract.

CSCS: Construction Skills Certification Scheme.

On site health and safety procedures

In this task you have to outline the procedures you would follow for either your current workplace or your centre to comply with health and safety requirements:

1. How would you summon emergency services to the site?

2. What information would emergency services require from you?

3. Describe how you would raise the alarm and evacuate the premises.

4. How would you recognise the correct escape routes?

5. What would you do if you saw a fire?

6. If a colleague has experienced an electric shock, what would you do?

key terms

Risks: situations that may arise, which could compromise your safety.

Risk assessments

Risk assessments are an important element of health and safety. Risk assessments are carried out to identify any potential **risks** to individuals. The purpose is to make sure that no one gets hurt or becomes ill due to work activity. It is a legal requirement for organisations to take a risk assessment of the workplace.

ACTIVITY

Outline in the following table the FOUR parts of the risk assessment you must complete for each of the given scenarios.

Identify the hazards	Who might be harmed and how?	What are you already doing?
Spot hazards by walking around workplace, asking co-workers what they think, checking manufacturer's instructions	*Identify groups of people who may be at risk. Some workers have particular needs, visitors and members of the public*	*List what is already in place to reduce the likelihood of harm or make harm less serious*
Manual handling Includes lifting, lowering, pushing, pulling and carrying of tools, materials and equipment		
Slips, trips and falls A facility is located underground, with access via concrete stairs leading to the basement Risk of falls from ladders when working at height Trip hazards from poorly stored materials and equipment		
Electrocution Risk from portable power tools Risk from working near live circuits		
Working with moving equipment Refurbishment work involves the movement of forklift trucks, cherry pickers and delivery vehicles		
Inadequate lighting The current underground car park has no external windows, therefore no natural light		
Exposure to asbestos Removal of old ceiling and wall partitioning Removal of flue systems/lagging		
Work-related stress Timescales for completing the refurbishment are tight, therefore long hours, overtime and shorter breaks		
Lack of communication between tradespeople Some workers may have poor English language skills Planning and scheduling of work without adequate consultation between trades		

There are five principles to risk assessments:

- identify hazards
- decide who might be harmed and how
- evaluate the risks and decide whether existing precautions are adequate
- record the findings of the assessment
- review the assessment and revise it, if necessary.

What further action is necessary?	How will you put the assessment into action?	Action by whom?	Action by when?	DONE (Please date)
What additional steps do you need to put in place to reduce potential risks?	*Action by whom, timescale and when completed. Remember to prioritise and deal with high risk hazards first*			

Who is responsible for health and safety?

ACTIVITY

In the following table, tick the people responsible for hazard and accident prevention. In some instances, more than one person may have a responsibility.

Hazard	Employers	Employees	Customer	Safety officer	HSE officer	Trade union rep	Environmental health officer
Working in poor light conditions							
Poor housekeeping of site							
Having a hangover							
Insufficient supervision							
Guards not provided around machinery							
Ladder with broken rungs							
Faulty PPE issued							
Fire exit blocked							
Wrong safety signs posted on walls							
Inadequate toilet facilities and access to drinking water							
Employer not informed of dangerous condition of building							
Mains voltage used on new building site							
Too much overtime working							
Accident book not kept up to date							
Death occurs on site due to a breach of Working at Height Regulations							

Accident reports

Every time there is an accident or injury on site, information about the incident must be logged in the accident book. Every company should have an accident book and you must ask your company where it is located if you are moving from site to site; usually this information will be given to you in the site induction, but if you don't know you must ask. You may also be asked to enter details on another casualty's behalf if they're not capable of doing so themselves. If the accident has involved a piece of faulty apparatus, you must not alter or interfere with it as it may be subject to an inspection from the Health and Safety Executive (HSE) if the accident was serious. Remember, the HSE will and can obtain entrance on site when required, without notice.

Complete this report form with a relevant scenario of your choice within the plumbing industry.

<table>
<tr><th colspan="2">PlumbBob Squared Pipes Ltd
Accident report form</th></tr>
<tr><td colspan="2">Full name of person who was injured:</td></tr>
<tr><td rowspan="2">Home address</td><td>Age</td></tr>
<tr><td>(please circle)
Male Female</td></tr>
<tr><td colspan="2">Status on location (please circle)
Employee Contractor Visitor</td></tr>
<tr><td>Date of accident:</td><td>Time of accident:</td></tr>
<tr><td colspan="2">Location of accident:</td></tr>
<tr><td colspan="2">What has caused the accident? (please give a detailed description)</td></tr>
<tr><td colspan="2">Names and addresses of witnesses</td></tr>
<tr><td colspan="2">Details of injuries</td></tr>
<tr><td colspan="2">Summary</td></tr>
</table>

First-aid treatments

ACTIVITY

Complete the table below. Decide the correct first-aid treatment and its effects for each problem listed.

First-aid treatment:	Effect:
Hold under cold running water	To reduce pain and swelling
Apply cold compress	To restore blood flow to the head
Position person with head between knees	To cool the skin
Lie person down and raise their feet	To remove object without damaging the eye
Wash out with clean water poured from a glass or a sterile eye-wash bath	To restore blood flow to the head, as head is in the lowest position and the feet are raised

Problem	First-aid treatment	Effect
Dizziness		
Fainting		
Minor burn		
Object in the eye		
Bruises		

If first aid is required, remember to get the help of a qualified first aider as soon as possible. If the person has had an accident, the details will need to be entered into the accident book.

key terms

PAT: portable appliance testing; a process in which electrical appliances are routinely checked for safety.

Competent person: someone who has been trained and certified to carry out the work needed.

Electricity safety on sites

The safe use of electricity on site is covered by the Electricity at Work Regulations, which came into force in 1990. Your employer is required to keep complete maintenance records of all portable equipment. All equipment must regularly pass a **PAT** test. When working on a site, everything from the microwave to your 110 V drill must pass this test completed by a **competent person** every three months and a PAT sticker must be put on the appliance after the test. If it is not PAT tested, you should not use it. Remember, when working on site you must reduce the voltage from 240 V to 110 V or even further by use of battery drills.

ACTIVITY

In pairs, discuss and make notes on the following. You may use the internet and other literature to help you.

1. You are in a domestic dwelling; how do you reduce the voltage from 240V to 110V?

2. What colour must all 110V equipment be?

3. Before using any machinery, what must be in place to help prevent accidents?

Working safely with heat producing equipment

Fire prevention

Fires could be the worst kind of hazard on a construction site. All fires should be investigated, however small.

Remember, there are three elements that need to be present for a fire to start:

- fuel – can be anything that will burn, e.g. wood, paper, flammable liquid
- oxygen – or air
- heat – a minimum temperature is needed but a naked flame, match or spark is sufficient to start a fire.

If you remove one of these elements, you will be able to put a fire out.

ACTIVITY

Fires can be prevented by employees taking care of their work environment. List SIX ways in which you could reduce the risk of fire on a construction site.

1.

2.

3.

4.

5.

6.

ACTIVITY

Fire fighting first actions

Listed below are FIVE steps to take in tackling a fire if you were the first person to discover it. Number the sentences in the correct sequence of steps you would take from 1 to 5.

_____ Evacuate the area

_____ Fight the fire (if you are trained to do so) but avoiding danger to life

_____ Raise the alarm

_____ Identify the appropriate fire extinguisher to use

_____ Close doors and windows to prevent the spread of the fire

Fire extinguishers

The type of fire extinguisher you would use varies depending on the nature of the fire. All fire extinguishers are colour coded so that you can see at a glance the type of extinguisher and what it contains.

British Standards BS 7863 means that a block of colour has now been placed above the operating instructions to cover 3–5 per cent of the extinguisher area. The most common types of fire extinguisher available are shown below.

> **key terms**
>
> **British Standards:** set standards of quality for goods and services, recognised by the award of a BS number. If something does not have a BS number, then it hasn't been approved.

KNOW YOUR FIRE EXTINGUISHERS
LABEL COLOUR CODES

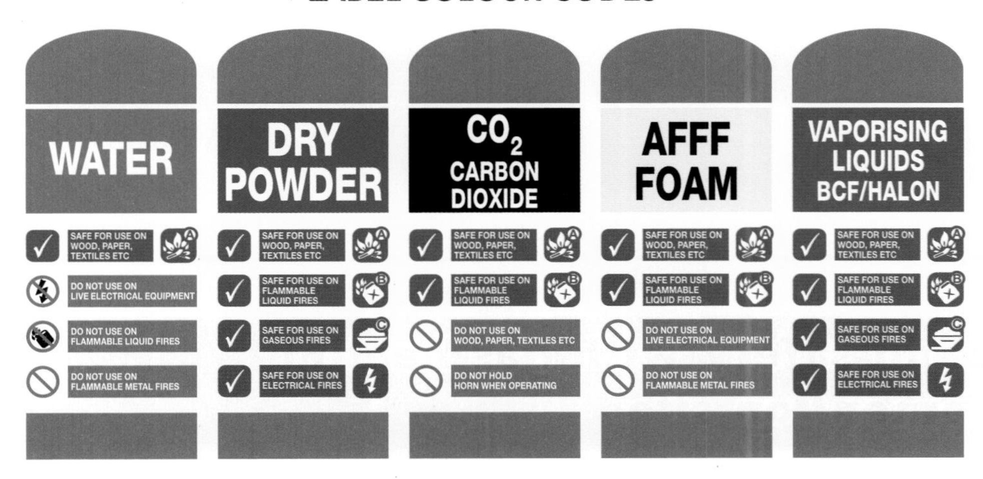

ACTIVITY

Identifying the correct fire extinguisher

Complete the table below indicating what product each fire extinguisher contains and which type of fire it should be used on.

Extinguisher label colour	Extinguisher contains:	Use this extinguisher on the following fire types:
White		
Blue		
Cream		
Black		
Green		
Yellow		

ACTIVITY

Go and find as many different fire extinguishers as you can in your workplace or your centre. Make a list of what you have found, where you found it and the type of fire it will put out. Share your findings with the rest of the class.

Working safely with gas heating equipment on site

key terms

Combustible: capable of catching fire and burning.

Permit: documentation issued for when you need to be given the permission to carry out an activity.

Once a fire is established it can spread quite quickly. Plumbing is a dangerous job and you can increase the fire hazards of sites. Blowtorches and welding equipment are frequently used near **combustible** materials, not taking into account the positions you need to work in, behind baths and in voids for instance.

You must obtain something called a 'hot works **permit**' to be able to complete the work you need to. This permit is critical because of the information it gives the site management.

Permit information

ACTIVITY

Which THREE pieces of information should you put on to the permit?

1.

2.

3.

Prevention

ACTIVITY

While you are completing works where there is a fire risk you should carry TWO vital pieces of equipment in your tool box. One will help you to prevent a fire and the other will help to put out a fire if something goes wrong. What are they?

1.

2.

Access equipment

Steps, ladders, roof ladders, hop ups, scaffolds and mobile scaffolds all fall under the term *access equipment* and are used for working at height. All access equipment should be regularly checked and its condition recorded. Sites now have a specific person who will do this and put a sticker on every apparatus weekly. However, you should also check the equipment before you use it as it may have been damaged.

ACTIVITY

Setting up access equipment

Listed below are EIGHT steps to setting up steps or ladders correctly. Number the sentences in the correct sequence you should take from 1 to 8:

____ The ground is level and clear of obstructions

____ Legs are firmly on the ground

____ You are facing your work

____ You have the correct footwear

____ Visually inspect steps for damage to rungs, welding and to check the feet are intact

____ Your hands are free to make three points of contact

____ The steps/ladders do not wobble when loaded with weight

____ There is a clear access for persons to walk around you; if not, erect barriers and signs

ACTIVITY

Safe use of ladders

In small groups, discuss why these people are using the ladders in unsafe ways.

1.

2.

3.

ACTIVITY

Hop ups are now more popular on site than old trestles. Use the internet to find out the HSE's policy on the use of this type of apparatus.

Using the information you have obtained, answer these questions with a yes or no.

1. Is it true that you can use a hop up on site?
2. Must you complete a risk assessment before use?
3. Can you use a hop up if you are working above 3 m?
4. Must you carry out a full inspection of the hop up before using?
5. Will the Working at Height Regulations 2005 give you more information and guidance regarding working with hop ups?

Working safely in excavations

When working on below-ground drainage you may be required to work in excavations. You must always ensure that you and others are safe. The main things you should always think about are the possibility of collapse of the excavation walls and the risk of persons falling in.

Remember, it is the responsibility of a competent person to check the excavations every day before use. It is your responsibility to keep yourself and others safe.

Wall collapse prevention

ACTIVITY

In small groups, discuss what you can do to prevent the walls of an excavation from collapsing. Use the internet to help you.

http://www.hse.gov.uk/construction/safetytopics/excavations.htm collapse

Write up your conclusions.

Fall prevention

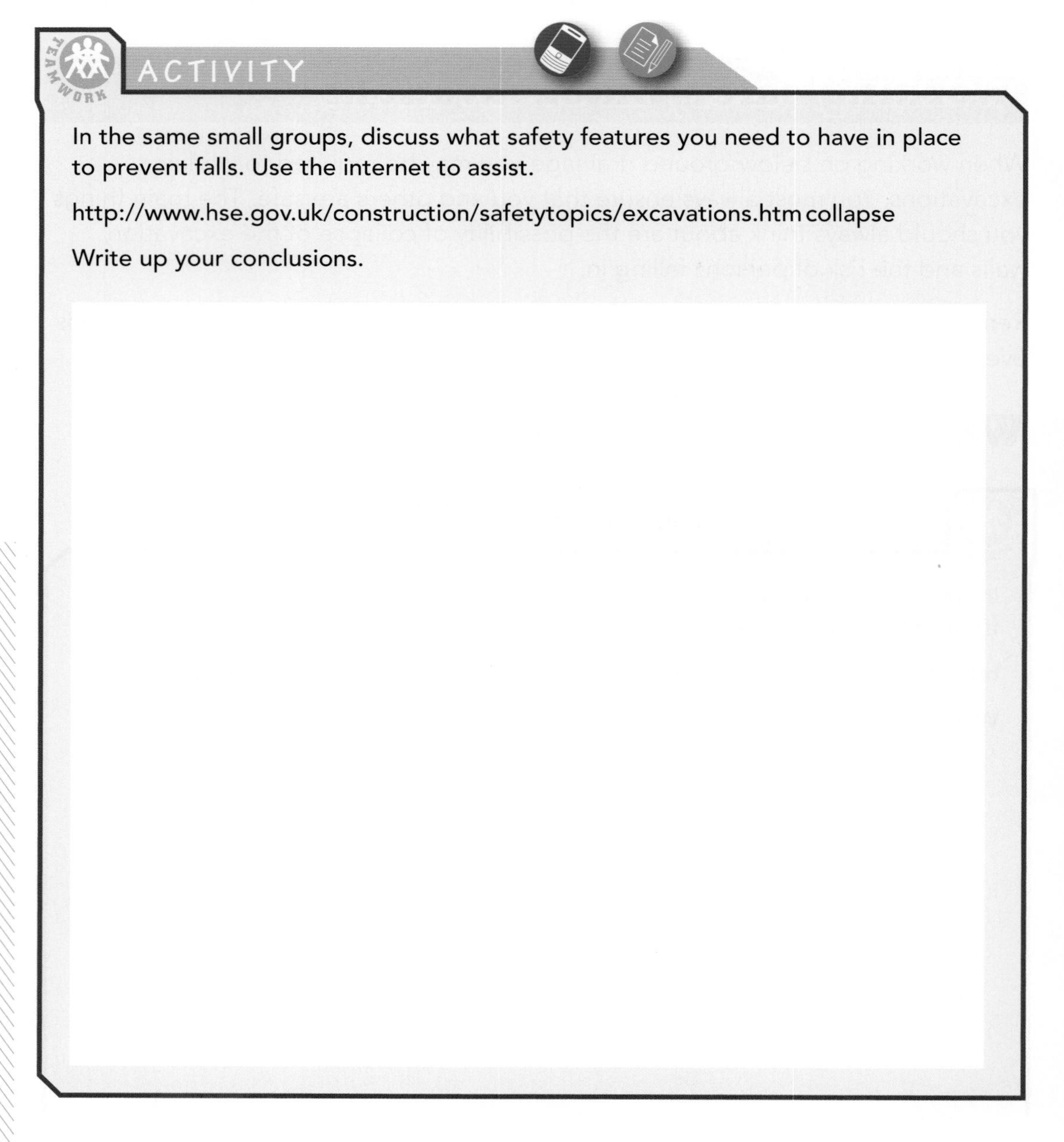

In the same small groups, discuss what safety features you need to have in place to prevent falls. Use the internet to assist.

http://www.hse.gov.uk/construction/safetytopics/excavations.htm collapse

Write up your conclusions.

Your questions answered...

I have been asked by my employer to remove a flue that looks like asbestos, what should I do?

There are strict rules about the safe removal of asbestos. You should not touch the flue until it has been tested to check what it is made of. Explain to your employer your concerns for not wanting to carry out the task and ask them to look at the flue so they can implement a test.

I need to use solvent cement for a longer time than stated on the data badge. Should I?

No. Always follow the guidelines given on the bottle and data badge. Have breaks and complete other work where you can and go back to what you were doing earlier. The guidelines may state that if the room is very well ventilated you may carry on. If you're unsure, ask your employer. If you start feeling dizzy or have head pains, then you should stop straight away and get some fresh air. You should then report this to your employer.

QUICK QUIZ

1. Who is required to have a site induction before they are allowed on site and who carries it out?
2. What type of identification proves you have health and safety awareness and knowledge?
3. What is the purpose of a risk assessment?
4. What does RIDDOR stand for? What does COSHH stand for?
5. What colours does asbestos occur in?
6. Which THREE elements are needed to create a fire?
7. Which type of extinguisher is best to put out an electrical fire?
8. What is the correct voltage to use at work?

Unit 002/202

Understand how to communicate with others within BSE

Clear communication between trades and personnel is essential for smooth working on site. Good communication skills come naturally to some, while others will find it hard. It doesn't matter if you are working on a multi-million pound site or in a domestic dwelling, talking to the architect or the customer, good communication skills will help you to deliver the works professionally and on time.

There are two main ways you might be employed as a plumber, either directly by the client, or on a sub-contract to a construction firm. Many of the thousands of plumbing businesses in the UK work for construction companies in this way. The main on site contractors will sub-let parts of their contract to specialists, such as electricians and plumbers. In whatever manner you are employed, you will need to effectively communicate with the people you come into contact with, from start to finish of every individual job. In order to do this effectively, you need to understand the roles of other members of the construction team.

key terms

Personnel: employees taken on by an organisation to carry out the works required.

Learning outcomes

- Know the members of the construction team and their role within the building services industry
- Know how to apply information sources in the building services industry
- Know how to communicate with others in the building services industry

Key knowledge

- Understand who you will be working with and what they do
- How to use the information others give you to assist completion of works
- Understand the importance of good relations within the workplace

key terms

Workplace: where you are working. This can be a construction site, but can mean other locations such as a customer's house, a school or a hospital.

Members of the construction team and their roles

Site personnel

Below is a wordsearch containing the types of personnel you are likely come into direct contact with in the **workplace.** Use the list below to assist you in this task.

Air con engineer Banks man Bricklayer Carpenter Cleaners Clerk Customer

Electrician Employer Foreman Heating engineer Labourers Merchants Painter

Plant operators Plasterer Roofer Site manager Tiler Ventilation engineer

C	R	E	L	I	T	E	A	G	O	N	S	L	A	B	O	U	R	E	R	S
X	D	B	M	U	M	S	J	E	W	N	I	E	J	P	C	K	G	Z	T	O
D	X	A	C	C	H	I	T	E	P	T	N	A	T	A	L	I	A	E	H	S
A	L	N	D	A	B	T	H	E	C	E	N	N	I	T	S	E	Z	L	G	R
W	H	K	N	S	H	E	A	T	I	N	G	E	N	G	I	N	E	E	R	E
N	A	S	O	G	D	M	B	D	B	Q	S	R	E	N	A	E	L	C	S	E
S	M	M	A	S	E	A	E	R	F	O	W	L	E	R	E	A	S	T	C	N
R	J	A	T	S	N	N	L	M	I	L	I	U	N	B	O	R	N	R	F	I
O	N	N	E	R	I	A	M	E	Q	C	U	S	T	O	M	E	R	I	L	G
T	A	T	A	E	S	G	U	C	Z	L	K	I	Q	M	A	Y	O	C	O	N
A	H	I	N	T	E	E	L	H	Z	E	I	L	W	N	S	A	E	I	O	E
R	G	A	T	N	R	R	L	G	D	R	L	F	A	L	A	N	N	A	P	N
E	O	L	S	I	O	R	E	X	X	K	I	M	S	Y	N	C	S	N	R	O
P	E	C	W	A	B	S	T	S	V	G	E	V	E	V	E	M	M	Y	E	C
O	C	A	R	P	E	N	T	E	R	R	Z	M	A	R	B	R	K	U	V	R
T	E	W	E	D	F	Y	T	J	A	T	D	V	P	J	C	L	H	U	I	I
N	N	I	F	G	B	M	N	F	Z	P	Q	L	B	L	T	H	J	J	L	A
A	O	N	O	C	L	F	O	R	E	M	A	N	I	A	O	G	A	S	J	K
L	V	V	O	V	F	G	E	N	Y	R	W	S	S	M	O	Y	W	N	D	C
P	R	E	R	E	T	S	A	L	P	H	U	G	O	S	Q	Z	E	Z	T	N
V	E	N	T	I	L	A	T	I	O	N	E	N	G	I	N	E	E	R	M	S

There may be some people on site that you see often, know who they are, but never get introduced to. This can be for various reasons and mainly due to the fact that your employer will have the main contact with these people.

Personnel – have you heard of these?

In groups, research any personnel from the wordsearch that you have not heard of. Discuss the elements of their job and how they might affect you as apprentice plumbers on site. You can do research on the internet or by asking your tutor or work colleagues. Make notes on your findings and your discussion.

ACTIVITY

In pairs, complete the table below. We have completed the first one for you.

Use the internet and other literature to help.

Operative	What they do
Building inspector	*Works on behalf of the local authority to ensure the works completed meet certain standards, for instance, that the building meets legal standards for access, and issue certificates for works completed*
Architect	
Structural engineer	
Clerk of works	
Quantity surveyor	
Building surveyor	

ACTIVITY

Workers on site

In pairs or individually, list the main trades that you will come into direct contact with because of your role.

Most of these people will be craft operatives. There are many other craft operatives on site who you will not have direct contact with every day. Can you think of THREE?

1.

2.

3.

ACTIVITY

Visitors on site

In pairs or individually, describe the main objectives of the following inspectors when they come on site. Use the internet to help you.

- HSE inspector
- Building control inspector
- Water inspector
- Electrical inspector

Communicating with others in the building services industry

In all business the use of good **communication** will make the work easier and help avoid problems. Something as simple as the way you ask somebody to pass a screwdriver can have a lasting effect on how you are **perceived** by others. Always be polite to other workers and clients, and remember your boss is depending on you to act professionally at all times when you are on duty.
There are three main ways to communicate with others:

- in writing
- verbally
- visually.

key terms

Communication: describes the various ways information is passed from one person to another, or to a group.

Perceived: how somebody will view you and your personality by the way you interact with them.

For each of the following situations, state what you think you should do.

- You have accidentally knocked over and broken an ornament in a customer's house:
- The customer is not happy with the run of pipework you have installed:
- A decorator has spilt paint on your pipe bender:
- The plasterer has asked you to put the water back on as you are holding her up:
- You need to order some soil pipes to finish a job:

key terms

Complaint: an expression of dissatisfaction with something.

Dealing with a complaint

In this task you should use a computer to write a letter to a customer after they have complained that your apprentice has left foot marks on the carpet entering the house. Your lecturer will acknowledge the letter as if they are the customer and will tell you whether or not they are happy with the outcome.

ACTIVITY

Visual aids for communicating

For each of the following, what are you being told?

1.

2.

3.

4.

5.

Employers' rights and responsibilities

There are three pieces of legislation relating to your employment:

- The Employment Rights Act 1996
- The Employment Relations Act 2004 (updated from 1999 Act)
- The Employment Act 2002

The Employment Rights Act 1996

ASK

ACTIVITY

Complete the table below. Fill in as much information as you can find using the internet and other literature.

The Act covers	Key points outlined
Wage rules	
Dispute resolution	
Dismissal and redundancy	
Time off work	

Employment Relations Act 2004 and the Employment Act 2002

ACTIVITY

Which one of these areas is not covered by this Act? Circle your answer.

Trade union recognition

The right to time off for union activities

Weekend leave

Industrial action ballots – procedures that must be adopted by unions

Maternity leave

Unfair discrimination when taking toilet breaks

Unfair dismissal of strikers

Improvements to work and tribunal procedures

Fixed term work directive

Parental leave

Time off for dependants – caring arrangements for close relatives

Employment tribunal awards – maximum levels, type of award

The right to be accompanied in disciplinary and grievance hearings

Work and parents

The right to be accompanied in meetings with your superiors

Part-time work – equality with full-time work in terms of pay, work conditions, holidays, etc.

Equal Pay Act

ACTIVITY

In small groups, discuss what the Equal Pay Act is trying to achieve. Make notes on your findings.

National Minimum Wage Act 1999

At the time of writing this book, there are three age ranges set out by the Low Pay Commission for different ranges of the minimum wage. Please be aware this may be subject to change, so if it affects you, always make sure you check the latest figures. It is your right to ask questions relating to pay and to find out how much you should be paid.

The three ranges are:

- 16–17 (but not apprentices whose wage structure is set by the JIB (Joint Industry Board))
- 18–21
- 22+.

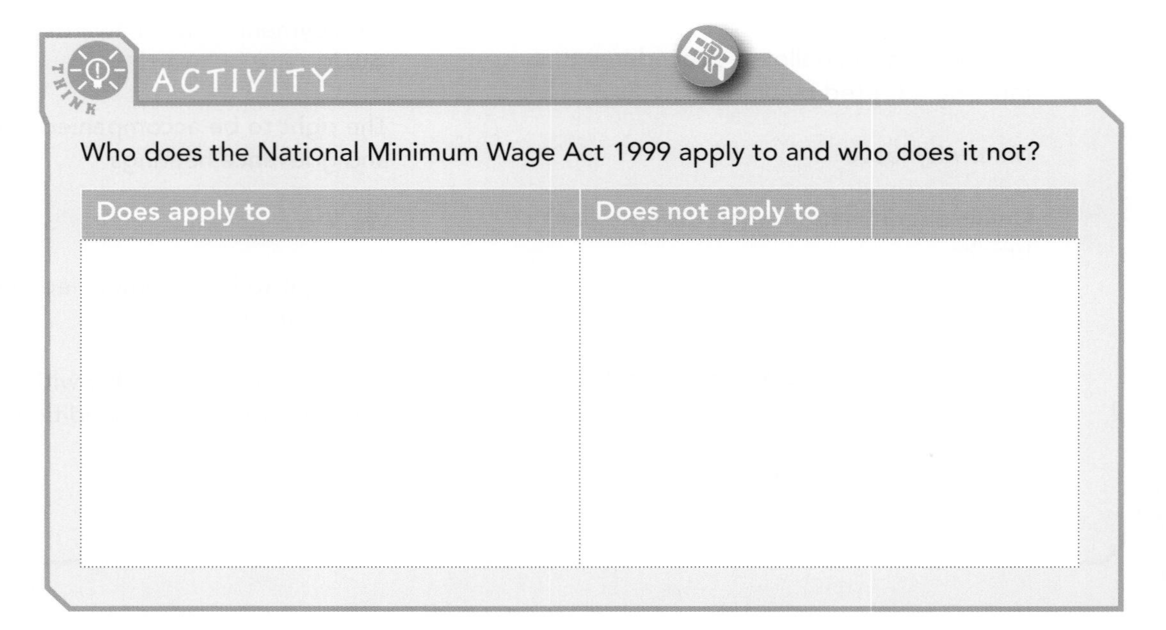

ACTIVITY

Who does the National Minimum Wage Act 1999 apply to and who does it not?

Does apply to	Does not apply to

Data Protection Act

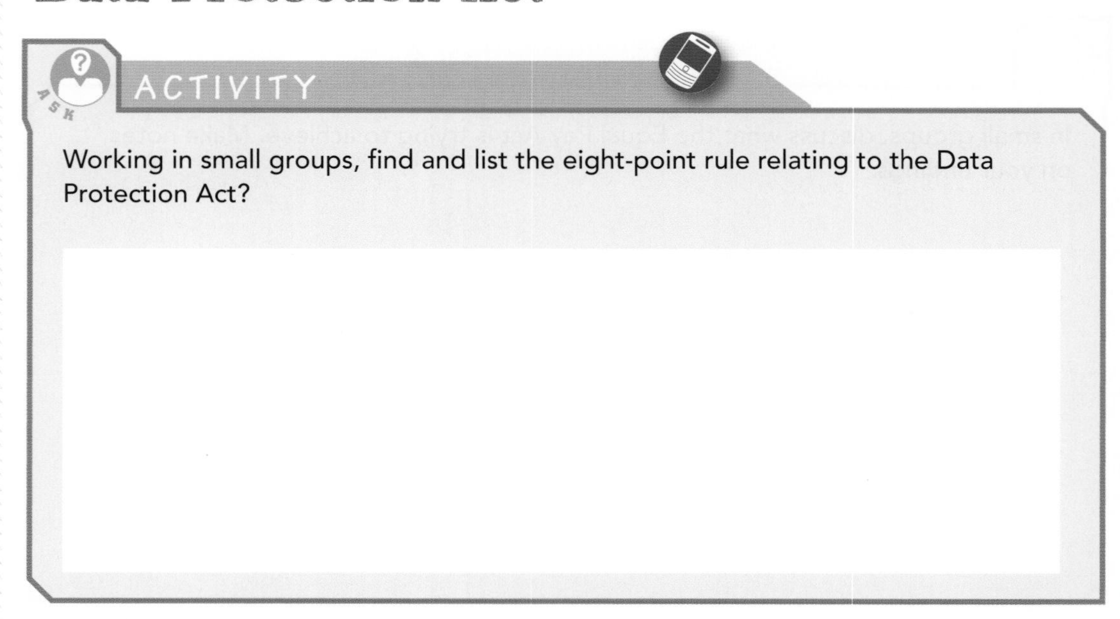

ACTIVITY

Working in small groups, find and list the eight-point rule relating to the Data Protection Act?

Careers

You need to be aware that there are a vast number of opportunities for you once you are qualified. You may wish to carry on working for someone or take the plunge and become self-employed. You may wish to see how far you can progress up a company. The great thing is that you are on your first step. Imagine a tree: you are the trunk, coming off are lots of branches, representing lots of career paths to explore. If one doesn't work out, choose another. You have many more options than if you never took on this qualification. Below is a diagram of some routes you may wish to follow.

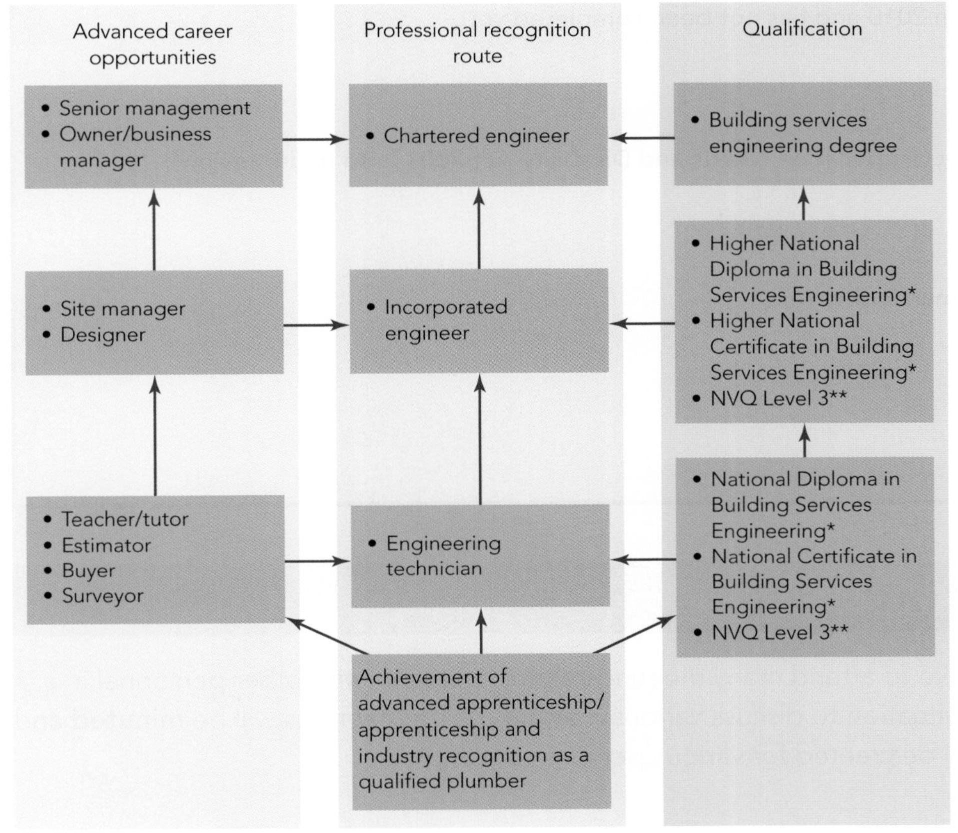

Notes
* These include a general Building Services Engineering (BSE) qualification, or optional routes such as Plumbing, Heating and Ventilating
** NVQ Level 3 qualifications currently exist in: Building Services, Engineering Technology and Project Management
Professional recognition is subject to approval by the Engineering Council

key terms

Equality: creating a fairer society, where everyone can participate and has the opportunity to fulfil their potential.

Diversity: when used as a contrast or addition to equality, it is about recognising individual as well as group differences, treating people as individuals, and placing positive value on diversity in the community and in the workforce.

Equality and diversity in the workplace

Throughout your working career, you will come into contact with various people from many different backgrounds and some people with disabilities. Attitudes towards **equality** and **diversity** in the workplace have improved vastly in the last 10 years and you must ensure that you fully understand your rights as well as those of others. For instance, when you are next on site working, think why there is always a downstairs toilet in new builds, or why the sockets are now set to a certain height.

Legislation reflecting equality and diversity

In pairs, produce a short statement outlining what aims each one of the following pieces of legislation is trying achieve.

1. Race Relations Act 1976 (Statutory Duties) Order 2001.
2. Disabilities Discrimination Act 1995 (be aware that this Act is under reform from 2010 and has not been completed yet).
3. Special Education Needs and Disability Act 2001 (also under review).
4. Human Rights Act.

key terms

Minutes: records of a meeting. A form of tracking who said what and what was said in a meeting and a record of action agreed.

Meetings

You will have to attend many meetings either on site or with other personnel in a domestic situation to discuss various matters. These meetings will be **minuted** and actions will be created for various persons to complete.

ACTIVITY

What are meetings intended to do?

In groups, discuss what meetings are for, how they work and their effectiveness.

Here is an example of some typical minutes. Read them through and try taking your own minutes at the next meeting you attend.

PlumbBob Squared Pipes Ltd

Meeting with the Plumbing Staff
Retro Lifestyle retirement site – Mayo, Belmullet

- 05/01/2012
- 4.30pm
- Induction room 1

Present – Bob Wilson (BW) Managing Director, Martin Willis (MW), Anthony Wang (AW), Hugo Wilson (HW), Amy Longland (AL)

Agenda

1. Progression of Plots 1, 2, 3, 4 and 5
2. Feedback sheets on completion
3. Work programme
4. Orders
5. Delivery notes
6. AOB

1. BW introduced the meeting agenda to the team and went through the progression of plots 1, 2, 3, 4 and 5. BW wants the apprentice (AL) to be more involved. HW has been instructed to mentor AL for the foreseeable future. All parties agreed.
2. MW highlighted a problem with the painters in plot 4 – BW to speak to the contractor to resolve.
3. MW handed over completion notes for the first plot to BW ready for the Clerk. MW expressed no concerns in the plot.
4. AW and MW to have a meeting with the foreman regarding the programme of works as they both feel plot 3 will be completed before time.
5. BW would like AL to be involved with ordering direct to the merchant to help with communication skills. AL highlighted she did not see the benefit of this but would complete as requested.
6. BW and MW have instructed AL and HW to take deliveries on site and pass over notes.
7. AOB –
   - Break times were discussed.
   - Smoking areas and where they have been moved to.
   - MW insisted all staff must wear PPE at all time on site or be removed by the foreman.

Action	Due date	Completion date
HW to mentor AL	05/01/12	Ongoing
BW to speak to painting contractors to resolve problem	05/01/12	06/01/12
AW to report back to BW after meeting with the foreman	07/01/12	
AL to start ordering materials directly	07/01/12	
AL and HW now in charge of deliveries	05/01/12	
ALL present to wear PPE at all times	05/01/12	
Next meeting 14/01/2012		

Applying information sources in the building services industry

Behind the scenes, and months before you start work on site, different people will be working on a range of documents relating to the work. You will need to understand these as you are very likely to come across all of them at some point. The information in these documents is very important and will help you with your everyday responsibilities.

They are:

- delivery notes
- building plans/drawings
- services plans/drawings
- work programmes
- job specifications
- manufacturers' instructions
- remittance advice.

ACTIVITY

Delivery notes

Give THREE reasons why delivery notes are important:

1.

2.

3.

PLUMBman Pat Ltd

Delivering to your workplace

PLUMBman Pat Ltd, Unit 5-6 The Heights, Binghamstown, Northampton, NN49 9EW
Tel: 01604 990011 Fax: 01604 999011

PlumbBob Squared Pipes Ltd
Pipedream Road
Semilong
Northampton
NN52 1JW

Account Number: 034571123
Statement date: 23/01/2011

Page number: 1
Served By: Hugo Wilson

Customer Ref	Placed By	Date Required	Order Type	PLUMBman No.	Delivery Address
PO33228	Bob Willis	16/03/2011	Delivery	77867	NN65 4PQ #21

Product Number	Description of goods	Quantity
PM93393	15 mm Yorkshire pipe	150 meters
PM8583	15 mm end feed elbows	25
PM32131	15 mm end feed tees	25
PM32455	15 mm end feed sockets	25

Received in good condition ....

Print Name ....

White copy – Customer Yellow copy – Branch

ACTIVITY

Building plan

Name FIVE pieces of information shown in the drawing below.

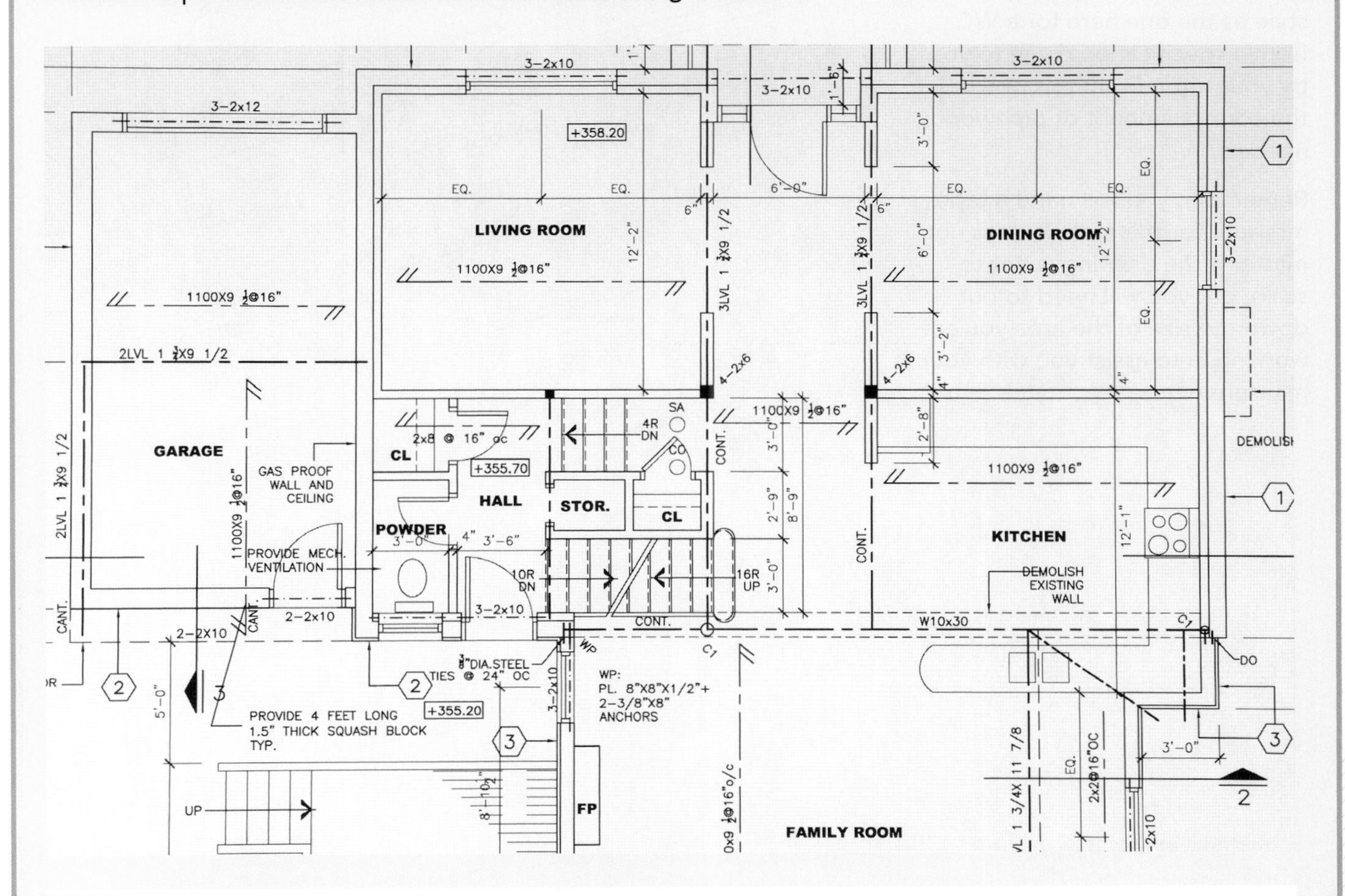

1.

2.

3.

4.

5.

ACTIVITY

Service plan

Complete a drawing in the same style as the one here for a WC from a specification given to you by your tutor. From this, work out the correct amount of pipework needed.

Remember, you will need a tape measure for this task. It does not matter if the drawing is not to scale, but you will need to put down the size of the area you are working in to assist you with your measurements.

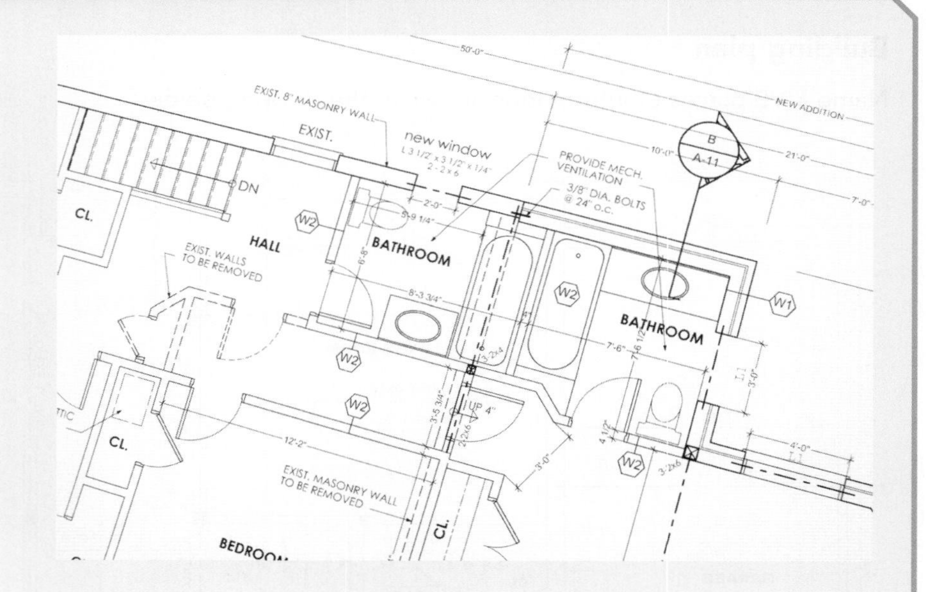

Amount of pipework and sizes required (show your workings)

Work programme

The work programme is vital to ensure you meet the finish date of a project.

Trade	Week 1	Week 2	Week 3	Week 4	Week 5	Week 6	Week 7	Week 8	Week 9	Week 10
Plumber										
Electrician										
Carpenter										
Plasterer										
Tiler										
Roofer										
Carpet layer										
Painter										
Gardener										

Above is a work programme for a refurbishment to a house in town.

In pairs, discuss what would happen to the programme of works if the following scenario happened:

The carpenter has been laying the floorboards after you have finished the pipe runs. When you pressurise the system you hear water running in every room. Then the lights go out. Obviously the carpenter has gone through your pipes, which you will now need to correct, and the water has damaged the lights and the ceiling.

ACTIVITY

Job specifications

State THREE pieces of information a job **specification** should give you:

1.

2.

3.

key terms

Specification: detailed document with an exact statement of particulars, especially a statement prescribing materials, dimensions and quality of work for something to be built, installed or manufactured.

ACTIVITY

There are other documents that will become more prominent later in your career. Some are listed below. Complete the table, using the internet and other literature to help you. We have completed the first row for you.

Document	What are they used for?	Who do you deliver the document to?	Communication styles
Estimates	*Estimates are rough quotes given for a certain job. Without going into full detail, they can give someone an idea of the cost of a job.*	*To a client*	*Verbally or draft letter/ written*
Quote			
Invoice			

Document	What are they used for?	Who do you deliver the document to?	Communication styles
Purchase order			
Time sheets			
Handover information			

Your questions answered...

I work on a major building site. I have not been given the drawings for the plot that I am working on and my boss has given me an oral explanation of where the pipe runs need to go. I have started work and noticed that there may be a mistake made by my boss regarding what he has told me. What should I do?

The first thing you should do is talk to your boss and explain your concerns. You should do this in a friendly manner as it may have been an oversight on their part, or you could be wrong. You should request the service drawings from the office as soon as possible. If you are correct, the other plots may also be completed incorrectly and, thinking back to the work plan, you can see how this might create a knock-on effect for other craftsperson's schedules. Once the outcome of the discussion has been agreed your boss will either instruct you to carry on with what you have been told to do, or speak with the foreman, clerk of works and the architect to decide on the best solution. Remember, all the sources of information mentioned in this unit are there to assist you in completing your job to the best of your ability, so ask to see them if you need to.

QUICK QUIZ

1. What is a sub-contractor?
2. Who is the banks man and what does he do?
3. Name another trade associated with the BSE sector?
4. How are meetings documented?
5. What is the purpose of the actions section from your meetings?
6. What is the purpose of a foreman?
7. If you are employed, what is the term for your status?
8. Which document will tell you the type of taps that should be fitted?
9. Which document will tell you your start and end date on a job?
10. If you need to find out where to put your appliances, which document would you use to gain this information?

Unit 003/203

Understand how to apply environmental protection measures within BSE

As a plumbing and heating engineer, you will be a member of one of the most important trades in the battle to help reduce the effect people have on the earth's resources. Take the UK as an example: there are many millions of households, all of which need water to drink and to wash; just where does that water come from? And why do we waste so much? If you think you don't waste water, just consider the act of brushing your teeth. Do you leave the tap on while you're brushing? If you do, you're not alone. If you do the maths to establish how much is wasted every morning by every person who leaves the tap on, you will probably be surprised. We waste millions of gallons of water in similar ways.

Everyone wants a nice warm house in the winter. To achieve this, you need to burn fossil fuels and the waste products from this combustion contribute to pollution and greenhouse gases in the atmosphere. Think about it. Every time you have a shower you are burning fuels; every time you turn on a light you are burning fuels.

In the UK, we have been working hard in improving the energy ratings for everything we use, from fridges and TVs to the fuel efficiency of our cars. But there are hundreds of thousands of homes out there that need to be brought up to a more energy efficient standard. For instance, every boiler you install will need to meet statutory energy efficiency standards and you will also be required to improve existing systems.

The disposal of waste is another aspect of environmental awareness that needs to be highlighted. If something can be recycled, then it should be. Unfortunately, the construction industry has been notoriously bad at recycling waste – that has to change, and you will be at the forefront of that change.

Learning outcomes

- Know the energy conservation legislation that applies to the building services industry
- Know the applications of energy sources used in the building services industry
- Know the importance of energy conservation when commissioning building services systems
- Know the methods of reducing waste and conserving energy while working in the building services industry
- Know how to safely dispose of materials used in the building services industry
- Know the methods of conserving and reducing wastage of water within the building services industry

Key knowledge

- Understand the legislation implemented to improve conservation of energy
- Know how to reduce waste and energy by the use of technology
- Understand the importance of safe disposal of certain materials

How to apply environmental legislation at work

Every industry will have its own regulations to follow and in the BSE sector we must follow the Building Regulations 2000, Part L1, which came into force in England and Wales on 1 April 2002. These regulations cover the mandatory minimum requirements and outline **best practice** to improve systems and assist in the conservation of energy and the reduction of waste products. Do not get too concerned if you don't fully understand all the systems available yet, as they will be explained in more depth in future units. This unit is intended to give you a head start in understanding how to apply environmental legislation across the systems.

key terms

Best practice: a statement describing the correct way to conduct a process.

ACTIVITY

For this task list below the mandatory requirements for Mechanical Sector in Part L. Use the internet to help you. Follow this link.

http://www.horstmann.co.uk/downloads/ElectronicDocuments/Literature/BuildingRegsPartL.pdf

ACTIVITY

Using the same document as for the activity on the previous page, input the FIVE requirements for Mechanical Sector in the left-hand column of the table below. You will need to break down the information they have given so you can interpret it more easily. You may need help, so ask your colleagues or your tutor for assistance. We have done the first one to show you how to complete the table.

Requirement title	What are the regulations asking you to do?
Hot water cylinders	Replace all cylinders with fully **insulated** BS 1566 as standard or BS 7206 for unvented systems. All cylinders need to have **thermostats** to control temperatures. If the heating system is also replaced, the system must be fully pumped.

key terms

Hot water cylinder: vessel that stores hot water for future use.

Insulation: material used to prevent or reduce the loss of energy from a surface.

Thermostat: device used for the control of temperatures.

Conservation: preservation of a commodity, such as oil or water.

ACTIVITY

Recognising controls that help with energy conservation within the home

In this task you should identify the controls illustrated below and explain their functions.

You can use the internet and/or books to help you.

1.

2.

3.

4.

5.

6.

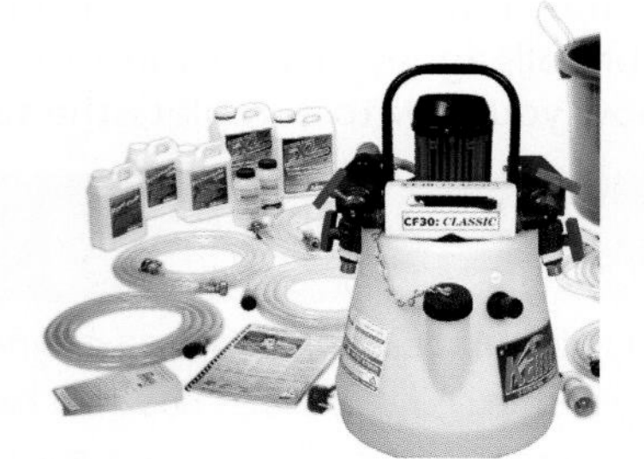

7.

ACTIVITY

In small groups, list as many other trades as possible that the Building Regulations (Part L 2002) will affect.

For each trade you have mentioned, give an example of how they help to reduce **emissions**.

key terms

Emissions: pollution or waste energy released to the atmosphere.

How to reduce waste and energy by the use of technology

SEDBUK and carbon output

Coal, peat, natural gas or **LPG** are high-carbon, non-renewable fuel types. SEDBUK ratings show the **efficiency** of burning these fuels. With the progression of technology we can now reduce our carbon output to a low level or even a zero carbon output.

key terms

LPG: liquid petroleum gas.

Efficiency: ability to produce a desired effect with minimum effort.

ACTIVITY

What does the abbreviation SEDBUK stand for?

Band	SEDBUK range
A	90+%
B	86% – 90%
C	82% – 86%
D	78% – 82%
E	74% – 78%
F	70% – 74%
G	60% – 70%

The image at the right shows the ratings of boilers. You must always install an 'A' rating boiler to ensure you are meeting the regulations.

ACTIVITY

Looking at the table below, you need to decide from the information given whether the system has a high, low or zero carbon output. We have done the first one to show you how. In groups, you should discuss the principles further to help your understanding.

Energy source	Carbon output (HIGH/LOW/ZERO)	Basic working principles
Solar thermal	LOW	Taking energy from the sun to heat water, which can then be pumped around the house providing heating and hot water
Natural gas		Burns natural gas to produce heat
Solid fuel (biomass)		Uses pellets, logs or chips to fuel a boiler in the same way as natural gas
Heat pump (water, air and ground source)		A machine or device that moves heat from one location (the 'source') at a lower temperature to another location. A heat pump can be used to provide heating or cooling. Even though the heat pump can heat, it still uses the same basic refrigeration cycle to do this
Combined heat and power (CHP)		Also known as a condensing boiler. It uses the 'lost heat' and recycles it to create a secondary source

Energy source	Carbon output (HIGH/LOW/ZERO)	Basic working principles
Combined cooling, heat and power (CCHP)		Excess heat produced is cooled by absorption chillers linked to the CHP system. This provides chilled water for cooling to be circulated around a building or community. This is particularly useful for schemes that require a large amount of air conditioning
Wind turbine		A wind turbine is a device that converts kinetic energy from the wind into mechanical energy. If the mechanical energy is used to produce electricity, the device may be called a wind generator or wind charger. If the mechanical energy is used to drive machinery, such as for grinding grain or pumping water, the device is called a windmill or wind pump
Solar photovoltaic		A method of generating electrical power by converting solar radiation into direct current electricity using semi-conductors that exhibit the photovoltaic effect. The method uses solar panels composed of a number of cells containing a photovoltaic material

Planning and commissioning systems

Your employer does not walk on site and start work without knowing exactly what has to be done. They will have carried out an assessment and worked out a systemic approach to the job so that they can maximise profit and prevent wastage.

Before you start any work, the way you plan your work activities will assist you in the conservation of materials.

After the completion of the installation, a full and precise commissioning procedure must take place to ensure that the system works to its optimum. This is the passing over of information to the customer about how the system works and when it should be serviced. This will help to ensure that the system works as efficiently as possible.

ACTIVITY

You are fitting a full heating system, but you haven't planned your order of materials or the daily activities correctly. Decide what the correct procedure should be and the effect that not doing something properly can have. We have done the first one as an example for you.

Heating system installation problems	Correct procedure	Effect of not doing the job correctly
Pipework runs not made	*Plan the exact route by investigating the runs and prepare the building fabric to suit*	*Overuse of plumbing materials by going the longer way round, including fittings and the gas used to solder the extra fittings. Using extra clips where not needed and notching extra joists, adding to the electricity used all round*

Heating system installation problems	Correct procedure	Effect of not doing the job correctly
Incorrect radiator sizes		
Running out of materials		
Not using TRVs or other controls		
Overcutting pipework		
Damage to the boiler as it wasn't stored properly		
Fittings have been stolen		

Once you have finished the installation, you will have another requirement of Part L to fulfil. This is to help the customer conserve energy and reduce their bills after you have gone by passing on the information about controls and how to operate the system correctly. In this way, you will be assisting the customer to reduce their effect on the environment.

ACTIVITY

You have just finished the installation of a full heating system and its controls. In pairs, discuss and make notes on the following:

1. **What documents should be passed over to the customer?**
2. **What demonstrations should you complete for the customer?**
3. **When should the system be maintained and serviced?**

Conserving and reducing water wastage

Water and Supply Regulations 1999 (WRAS)

key terms

Contamination: the mixing of substances of which one could cause harm.

The Water Regulations are set to assist in the prevention of waste, undue consumption, misuse or **contamination** of water. Owners and occupiers of premises and anyone who installs plumbing systems or water fittings will have the legal duty to ensure that the systems are up to standard. In most cases, advanced notice must be given of proposed installations so architects, building developers and plumbers have to follow these regulations on behalf of future occupiers.

REFLECT

ACTIVITY

Describe THREE situations that could involve water being wasted. Remember the example of leaving a tap running whilst brushing your teeth.

1.

2.

3.

Many large businesses are taking the conservation of water very seriously and investing in the technology available to reduce its use. In the end, this investment will save thousands of pounds. For instance, next time you are in a public lavatory look out for push button taps that turn off by themselves and infrared flushes. These small improvements potentially save thousands of gallons of water.

ACTIVITY

Look for as many pieces of technology used to prevent the waste of water that you might come across in your daily routine as you can.

Don't forget your **cistern** at home: you may have a two-button dual-flushing device; but it doesn't stop there, as all cisterns have a water mark to which the water should be set. There are also valves that limit the speed of water or backflow to stop contamination. It doesn't matter if you don't know the correct name of the piece of technology at this stage, but now is the time to start finding out.

What you have seen	Correct name	What does it do?

key terms

Cistern: a receptacle for holding water or other liquid.

Rainwater harvesting

Rainwater harvesting is the accumulating and storage of rainwater for reuse before it returns to the ground. Rainwater from roof tops can be harmful to human health but can be useful to flush toilets, wash clothes, water the garden and wash cars; these uses alone can halve the amount of water used by a typical home.

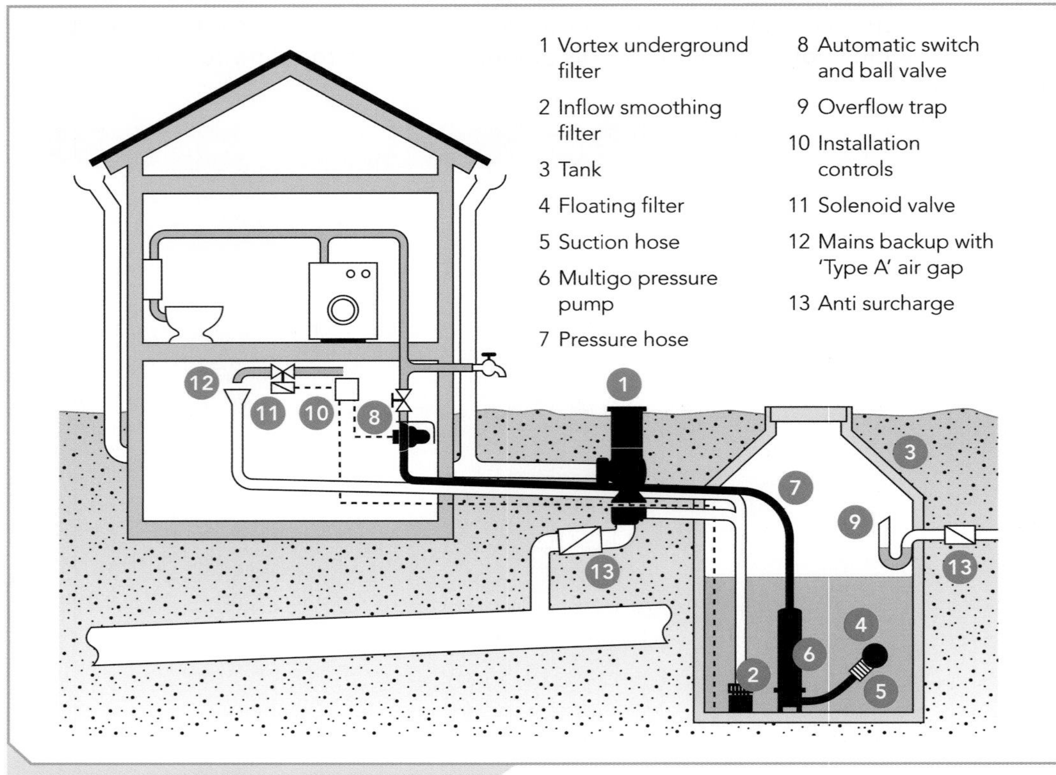

A basic ground source water system

ACTIVITY

The rate at which water can be collected will be dependent on the plan area of the system and the size of the roof. In some areas, the intensity of rainfall will affect the recovery time of the system.

To work out how much a system could collect you will need to do some maths. You will have to find out some information first, such as:

- annual **precipitation** (mm per annum) × catchment area (roof) m^2

This will equal litres per annum yield.

So, a 100 m^2 roof catchment catching 1,000 mm/p.a. yields 100 kL/p.a., which is a vast amount.

Show your workings for the following:

- you have a roof that is 300 m^2
- annual precipitation is 500 mm

key terms

Precipitation: a deposit on the earth of hail, mist, rain, sleet or snow; also, the quantity of water deposited.

What is the yearly yield of water that may be harvested?

How to safely dispose of materials

The most important piece of legislation that covers the removal of waste products from the working environment is the Site Waste Management Plans Regulations 2008. These regulations require a site waste management plan to be drawn up and implemented by contractors for all construction projects with an estimated cost greater than £300,000. The plans must record detailed estimates of the types and quantities of waste that will be produced, and confirmation of the actual waste types generated and how they have been managed. Failure to produce or implement a plan is punishable by a fine of upwards of £50,000.

The construction industry is one of the biggest users of building materials in the UK. At present, the industry uses about 400 million tonnes of materials per year, but some 100 million tonnes ends up as waste. Not all the waste is correctly disposed of either. The illegal disposal of waste is called fly tipping.

ACTIVITY

To fill in this table, you need to decide how you would dispose of the materials listed in the left-hand column and what might happen if you did not do this correctly. Working in pairs or small groups, carry out some research using the internet and written materials.

Material	Correct disposal method	Environmental effect of not disposing properly
Asbestos		
Metals		
Plastics		

Material	Correct disposal method	Environmental effect of not disposing properly
Wood		
Cardboard		
Electrical/electronic equipment		
Refrigerants		
Solvents		
Acids/heavy duty cleaning products		
Gas cylinders		

Your questions answered...

We have fitted a new heating system in a dwelling and, to cut costs, my boss has not fitted TRVs and a room stat. Is this OK?

The short answer is 'No'. In this unit you have been asked to look at a document called Part L, which states that all systems that have been altered or fitted new should consist of a certain amount of specific fittings and controls designed to increase efficiency and conserve energy.

TRVs (thermostatic radiator valves) are fitted to control the temperature of individual rooms. This results in the customer having a choice of which rooms require the 'most' heating. If they have a spare room that no one uses, then they have the choice of shutting down that radiator altogether and they may wish to limit the temperature in further rooms. This conserves energy and reduces waste.

The room thermostat, which should be fitted in an area that will give the ambient temperature of the house, is connected directly to the heat source – the boiler. Again, the customer can control the temperature of the house as it suits them. Once that temperature is achieved, the thermostat will tell the heat source to turn off.

If these two controls are not fitted, the heat source will only turn off when it reaches the maximum temperature requested on its own thermostat. It will carry on stopping and starting every time it reduces in heat. This would lead to a massive increase in the ratio of fuels used compared to what is really required for the customer's comfort.

Part L is a requirement that must be followed to bring systems up to standard and help reduce wastage.

QUICK QUIZ

1. What does Part L cover?
2. What are regulations for?
3. What do the letters SEDBUK stand for?
4. What rating should all boilers have?
5. There are THREE carbon ratings, what are they?
6. What is the overall purpose of explaining controls to the customer/client?
7. Name all the thermostats that must be on systems to comply with Part L.
8. How does insulating pipes help to reduce wastage?
9. What does the abbreviation WRAS stand for?
10. How many tonnes of building materials do we use in the UK on average in a year?

Unit 004/204

Understand how to apply scientific principles within mechanical services engineering

Ideally, you should understand the fundamental workings of all systems within the sector and have knowledge of the basic maths and science relating to these. This knowledge will help to speed up installations and reduce mistakes throughout your projects, along with giving you an understanding of how something works and why.

You will be judged not only on your practical skills but also on your competency once the system has been installed – there is no point installing something if you don't understand how it works. Sounds scary? Well yes, but once you understand the basics you will gain confidence and see how important it all is. Ask yourself this: how much water does it take to fill your bath? Not only is this important for the size of a **CWSC** you need to fit in a dwelling, but also to the recovery of the system itself. What restricts this recovery and why does it occur? You need to be able to answer such questions to show that you are competent in these basic skills.

We recommend going through this unit at least twice until you are confident of your knowledge.

key terms

CWSC: cold water storage cistern.

Learning outcomes

- Know the standard units of measurement used in the mechanical services industry
- Know the properties of materials used in the mechanical services industry
- Know the relationship between energy, heat and power in the mechanical services industry
- Know the principles of force and pressure and their application in the mechanical services industry
- Know simple mechanical principles and their application in the mechanical services industry
- Know the principles of electricity as they relate to the mechanical services industry

Key knowledge

- SI units of measurement within the mechanical sector
- Area, volume and capacity and where they are used within the industry
- Properties of water and its relationships within plumbing
- Density of water, materials and gases
- Boyle's law and Charles's law
- Pressure and force of substances
- Siphonage and capillary action
- Materials and their properties, including corrosion
- Properties of heat and how the energy is transferred
- Basic electrical principles and earth bonding
- Series and parallel circuits

Standard units of measurement

SI units

When engineers, scientists and technicians get together to work on design and construction projects, they need to use a common language or form of expression to understand each other. So before we are able to communicate in a mechanical sense, we must first become familiar with the units of measurement and know how to use those units in a mathematical manner.

In the UK and many other European countries an international system of units is used. This is known as Système Internationale (SI) units. Each of these units has a symbol to represent it, but these should not be confused with the unit names. These will become more apparent and make more sense as they are used in different formulae and equations.

ACTIVITY

The following table lists the SI units you need to recognise and identify. Complete the table, putting in the correct symbol, name and definition of the quantity in your own words. The first row has been completed for you.

Quantity	Symbol	Name	Definition
Length	*m*	*Metre*	*The linear extent from one end to the other*
Area			
Volume			
Mass			
Density			
Time			
Temperature			
Velocity			
Capacity			
Pressure			
Force			

key terms

Velocity: the speed and direction of motion of a substance flowing from one point to another.

Mechanics: the branch of physics concerned with the working parts of a machine.

Basic mechanical principles

Mechanics is all about *machines* and how we use them to make life easier.

A simple machine can be defined as a device that helps us to perform our work more easily when a force is applied on it. Machines also allow us to use a smaller force to overcome a larger force and to help us change the direction of the force and work at a faster speed.

Screws, wheels, axles and levers are all simple machines.

ACTIVITY

Leavers, gears and pulleys are all types of 'machines'. Find day-to-day examples of these machines, then draw a diagram of them in the space provided below. Write in your own words how they work.

1. Levers

2. Gears

3. Pulleys

Area, volume and capacity

You will need to understand the terms *area*, *volume* and *capacity* when designing and installing plumbing systems.

Area

Area is the flat measurement of a space, calculated from a plan or from actual linear measurements taken on the ground. It is expressed in metres squared (m^2). This is known as the plane area. The formula we use for measuring the area of a simple room shape, such as a square or a rectangle, is very straightforward:

length × width

The calculation of the area of more complex shapes, such as irregular shapes and circles or cylinders, is more involved, however, and there will be more about this later.

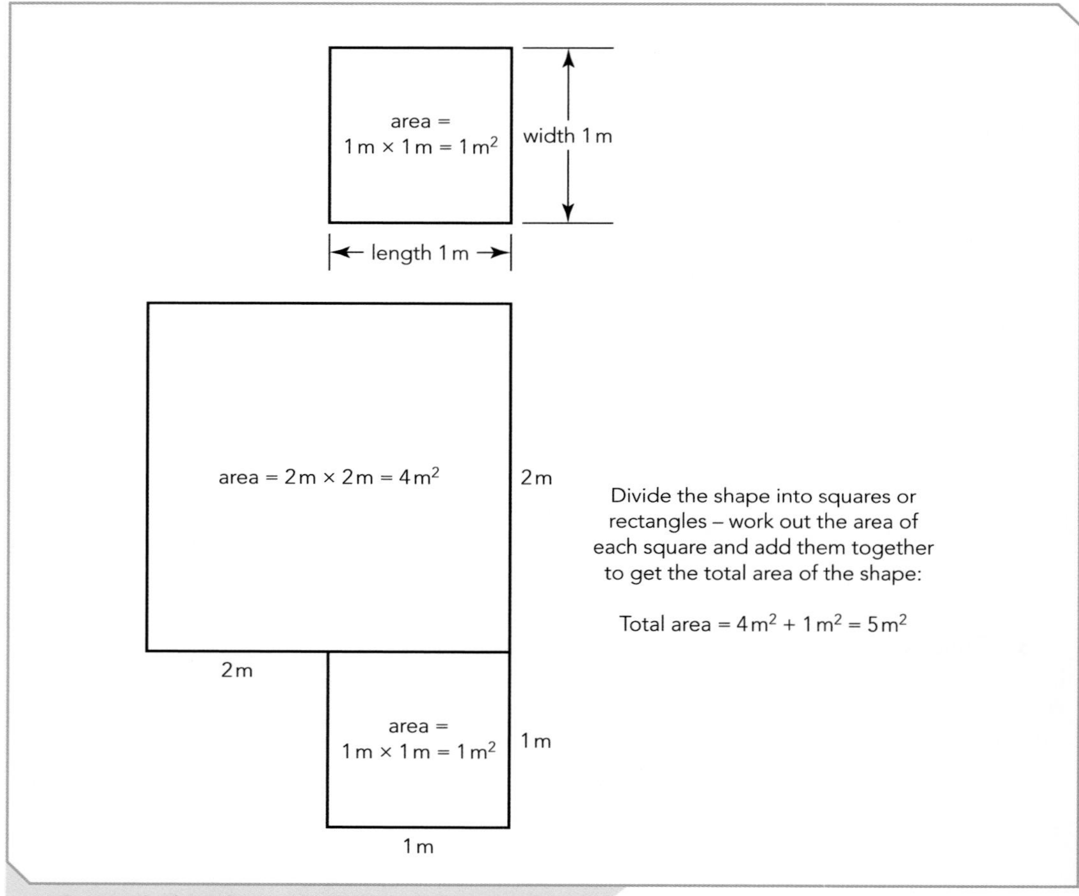

How to calculate the area of a square or rectangle

Volume

Let's move on from flat or plane shapes to three-dimensional (3D) shapes.

A 3D square or rectangle is called a cuboid. You can think about cuboids simply as 'boxes'.

The amount of space enclosed by a 3D shape is called the volume. The volume of cuboids can be calculated by using the simple formula:

length × height × width

Volume is measured in cubed units, e.g. mm^3.

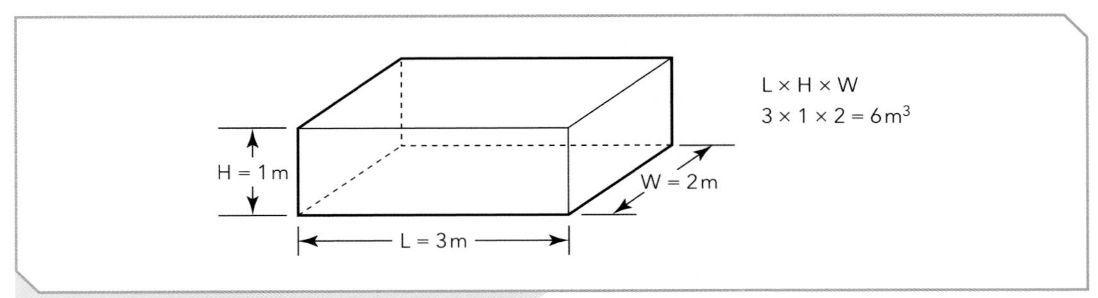

How to calculate the volume of a cuboid

In plumbing the volume of a vessel is expressed as its 'capacity', which tells us the maximum amount of matter it can hold, for instance the maximum amount of water that can be held by a CWSC or hot water cylinder.

Capacity is measured in litres, and this can be simply calculated by multiplying the volume in m^3 by 1,000.

For instance, if the volume of a CWSC is $1.25\,m^3$, its capacity will be $1.25\,m^3 \times 1{,}000$

= 1,250 litres

Cylinder calculations

To make cylinder calculations, first you need to know about radius and diameter.

The *radius* is the distance from the centre of the cylinder to its edge.

The *diameter* starts at one side of the circle, goes through the centre and ends on the other side.

So the diameter is twice the radius:

diameter = 2 × radius

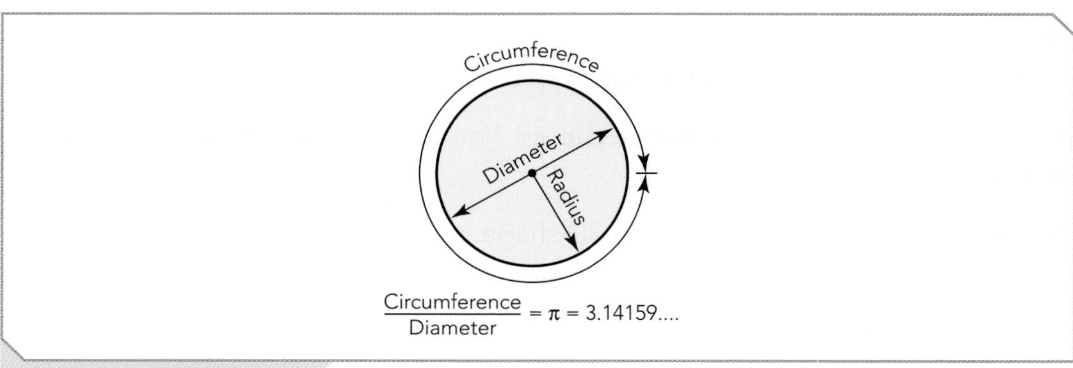

Dimensions of a circle

key terms

Pi: a number that represents the ratio of the circumference of any circle to its diameter. The value of pi is 22/7 or 3.142 and this number is used in calculations to find out the dimensions of circles, spheres and cylinders.

The *circumference* is the distance around the edge of the circle.

It is exactly **pi** (the symbol is π) times the diameter, so if:

circumference = π × diameter

these are also true:

circumference = 2 × π × radius

circumference/diameter = π

And the volume of a cylinder would be:

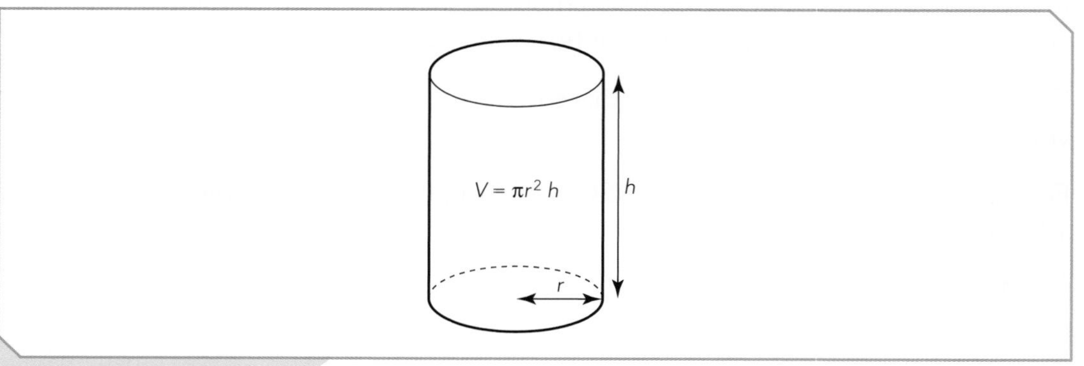

Volume of a cylinder

ACTIVITY

For the following task you will be required to show your workings alongside each drawing.

Question 1

What is the overall *area* of the space below?

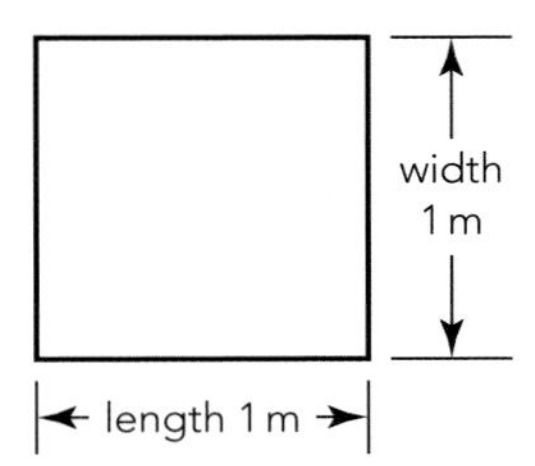

Area of a square

Question 2

What is the *volume* of the cuboid below?

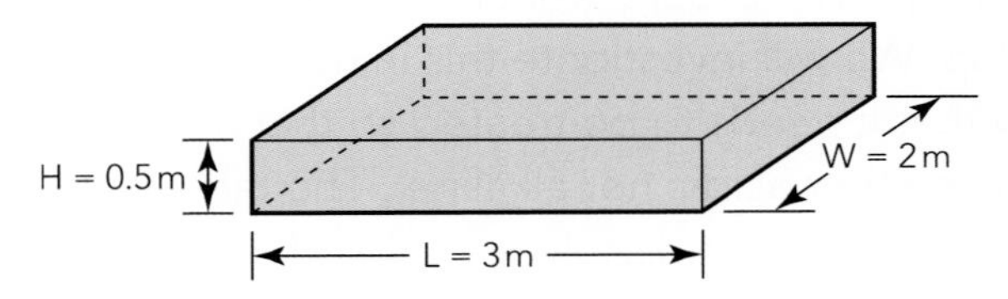

Volume of a cuboid

Question 3

What is the *capacity* of the cylinder below?

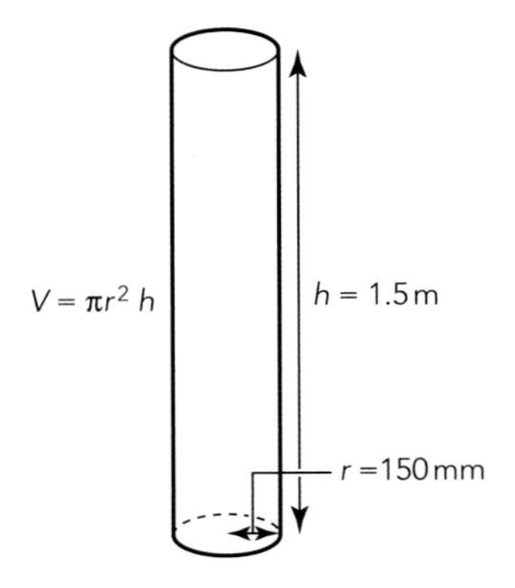

Capacity of a cylinder

Properties of water and its effect on different materials

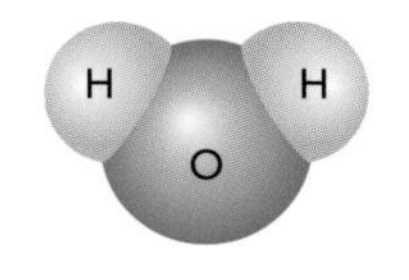

A water molecule, H_2O

We all know that without water we would die. We have also talked about it being a commodity that is wasted and some of the measures we take to reduce that waste. But what is water? Chemically speaking, it is two atoms of hydrogen and one atom of oxygen, which can be expressed as H_2O.

Water is also a solvent. This means it is able to dissolve a lot of materials and keep them in solution until a change in the environment causes them to come out of solution (a process called *precipitation*). This is what happens with limescale. Calcium carbonate is dissolved in the water when it passes through rocks, such as chalk or limestone, and comes out of solution when the water evaporates, for instance when the system temperature is increased. You will be familiar with limescale being deposited in kettles, hot water taps and on the walls of a hot water cylinder or heating pipes.

pH values and water hardness

key terms

pH: power of hydrogen – a measurement of the acidity or alkalinity of fluids.

The **pH** value of water is a measure of how acidic or alkaline it is. Before you take that first drink of water in the morning, think exactly what has happened to that water before it has reached your tap. We will investigate this in more detail later on, but for now you need to know that it needs to be treated so that it has the correct pH value – close to neutral (neither acidic nor alkaline). The table below shows what happens to the pH of water when it dissolves or absorbs different materials.

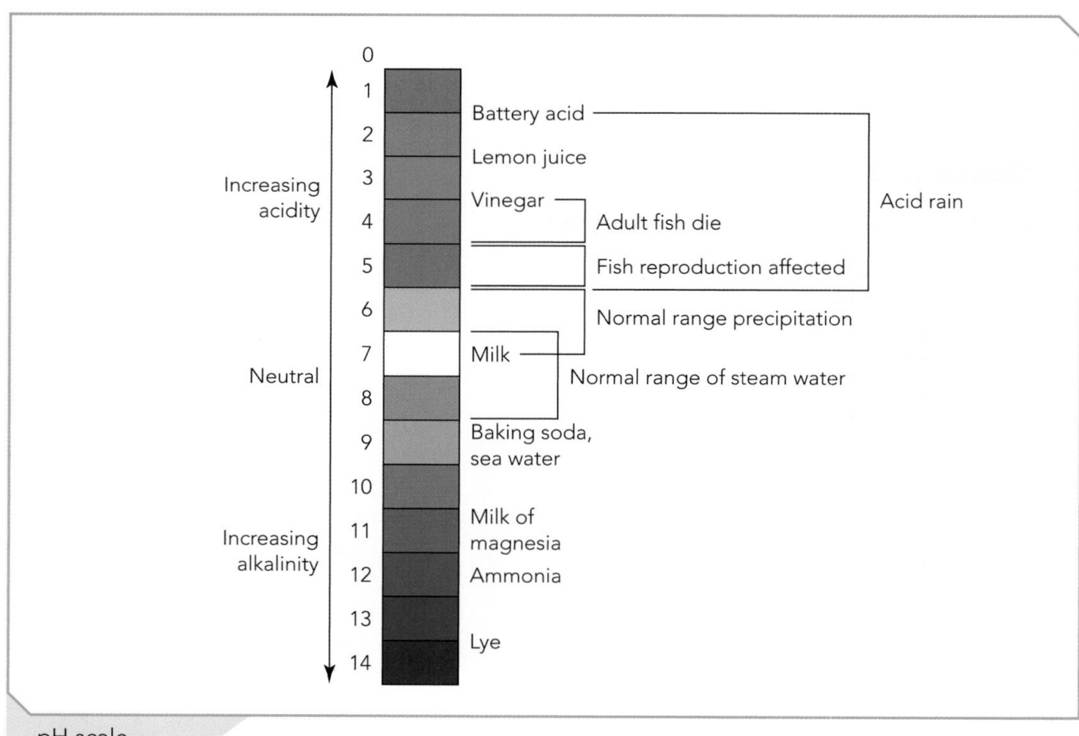

pH scale

ACTIVITY

1. What is the pH value of pure water?

2. Why would this change for normal 'tap' water?

3. What will happen to the pH value if sulphuric acid is added to water?

4. What will happen to the pH value if water dissolves limestone?

ACTIVITY

Acidic and alkaline water

In groups, discuss the various places where you as a plumber will find water with a large amount of alkalinity or acidity. Make notes on your discussion.

ACTIVITY

Water hardness

From the last task and the notes you made, and from new research if necessary, in your groups answer the following questions:

1. What is soft water?

2. What is temporary hardness?

3. What is permanent hardness?

ACTIVITY

Properties of water passing through different ground strata

Complete the table below by yourself and discuss your findings in your group afterwards. Use the final column to make notes of other people's findings.

Type of ground condition	What is being added to the water as it passes through?	Will the water be hard or soft? Acidic or alkaline?	Notes:
Salt			
Peat/moorland			
Chalk			
Sandstone			

Properties of water in its different states

Water does not always come in liquid form. Sometimes it is solid (ice), sometimes a gas (water vapour or steam). Different terminology is used for the different processes in which water changes from one state to another, such as melting, freezing, evaporation and condensation. When water changes from one state to another it is changing its structure – that is the way its molecules are arranged.

Take condensation as an example. Condensation occurs when the water molecules that exist as water vapour in hot air touch a cold material and form droplets of liquid water. There are natural points when these changes occur. Heat produced or absorbed during these processes is known as *latent* heat. By contrast, *sensible* heat refers to energy that causes a change in temperature. Latent heat does not change the temperature, only the state of the water, whereas sensible heat can be felt (sensed) as heating or cooling.

ACTIVITY

Properties of water in different states

Complete the table below.

H_2O and known form	Term used by us?	Explain what is happening to the molecules in this state?
Liquid		
Gas		
Solid		

Density

When a substance changes state, the **density** of the substance also changes. As a mechanical engineer, it is important that you understand that the density of water changes with temperature. Water is less dense when it's heated; its molecules start to move away from each other:

$1\,m^3$ of water at 4°C has a mass of 1,000 kg
$1\,m^3$ of water at 82°C has a mass of 967 kg

Other mechanical materials have their own density. We work out density by using the following formula:

$$\text{density} = \frac{\textbf{mass}}{\textbf{volume}}$$

key terms

Density: the degree of compactness of the molecules of a substance.

Mass: not to be confused with weight, mass refers to the quantity of matter a substance contains.

Volume: the amount of three-dimensional space an object occupies.

ACTIVITY

Mass and weight

Answer the questions below. Use as many sources as you need to find out the answers.

1. What is the symbol for the SI unit of mass?
2. What is mass measured in?
3. What is weight measured in?
4. What is the relationship between mass, weight and gravity?

TEAMWORK

ACTIVITY

Working in pairs, use the internet to help you complete the table below. You know the relative density of water is 1; we have filled this one in for you.

Material	Relative density
Water	1
Copper	
Low-carbon steel	
Lead	
Brass	
Plastic	

REFLECT

ACTIVITY

Summarise what is meant by the term 'relative density'. Think of the relationship between water and copper.

Gases

Density of gases

In the mechanical engineering sector, you will work with and therefore need to understand about gases and their different densities. Before we talked about density with relation to water, with water having the value of 1. With gases, we talk about the density of gases relative to air, where air has a relative density of 1.

ASK

ACTIVITY

Use the internet to help you complete the table below. Air has a relative density of 1; we have filled this one in for you.

Material	Relative density	Is this gas lighter or heaver than air? Will the gas rise or fall to the floor?
Air	1	
Natural gas		
Butane		
Propane		

Material	Relative density	Is this gas lighter or heaver than air? Will the gas rise or fall to the floor?
Carbon dioxide		
LPG		
Refrigerant gases		

Boyle's law

In the mid-1600s, Robert Boyle studied the relationship between the pressure (p) and the volume (V) of a confined gas held at a **constant** temperature. Put in simple terms, the conclusion that Boyle came up with was that when pressure is increased the volume will decrease. Therefore, if the pressure is decreased the volume will increase.

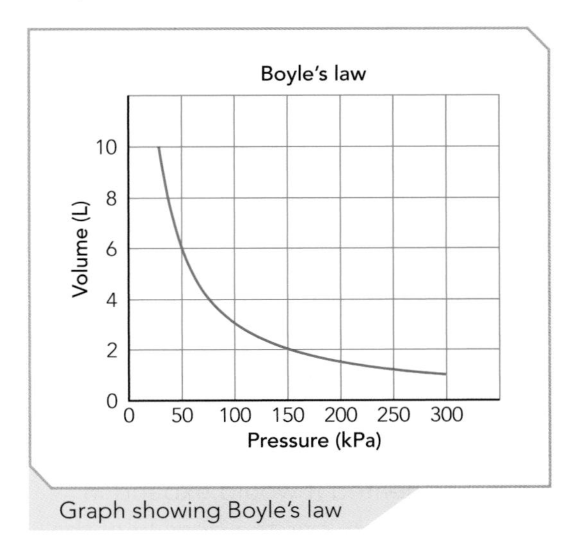

Graph showing Boyle's law

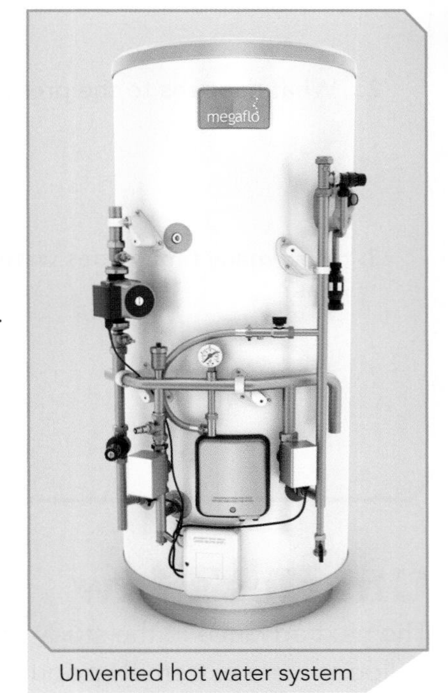

Unvented hot water system

key terms

Constant: permanent or unchanging. Often applied to a mathematical value that does not change.

Vented: when a system is not sealed and is classed as being atmospheric, working under normal conditions.

The equation is:

$pV = k$

where k = a **constant** value representative of the pressure and volume of the system.

LPG bottle

Boyle's law has a real impact on how our systems work in the industry. LPG, for instance, is a gas compressed into a liquid. As a liquid it has a much smaller volume and can then be easily stored under pressure in bottles. This has a direct relation with Charles's law, which we will look at below.

Boyle's law has also been applied to the unvented cylinder system (pictured above). Water at normal atmospheric pressure will boil at 100 °C. In plumbing terms, a **vented** hot water system will work at atmospheric pressure. If you increase the pressure or pressurise the system by sealing it, water or gas will have nowhere to go. The pressure will increase, therefore increasing the boiling point. Keeping the temperature under control so it does not turn to steam will keep the system safe.

ACTIVITY

For each of the following questions, write a short paragraph of explanation with reference to Boyle's law.

1. Using Boyle's law, explain what happens to the liquid in a sealed system when it is pressurised.

2. What happens to the pressure in a sealed system if heat is introduced?

3. How many times does water volume increase when it turns to steam or a gas?

Charles's law

The next advance in the study of gases came in the early 1800s in France. Hot air balloons were extremely popular at that time and scientists were eager to improve the performance of their balloons. Two French scientists, Jacques Charles and Joseph-Louis Gay-Lussac, made detailed measurements on how the volume of a gas was affected by temperature. When the gas was heated it would expand in the balloon, lifting it as the density of the warmed gas decreased.

So what if the temperature is decreased? This will result in a decrease in volume. Gases become liquids when the temperature is decreased enough. This is called liquefaction. The graph does not show this but instead shows how the volume of the gas would theoretically decrease at low temperatures if there were no liquefaction. An important thing to remember is that the total number of molecules is constant and does not increase or decrease.

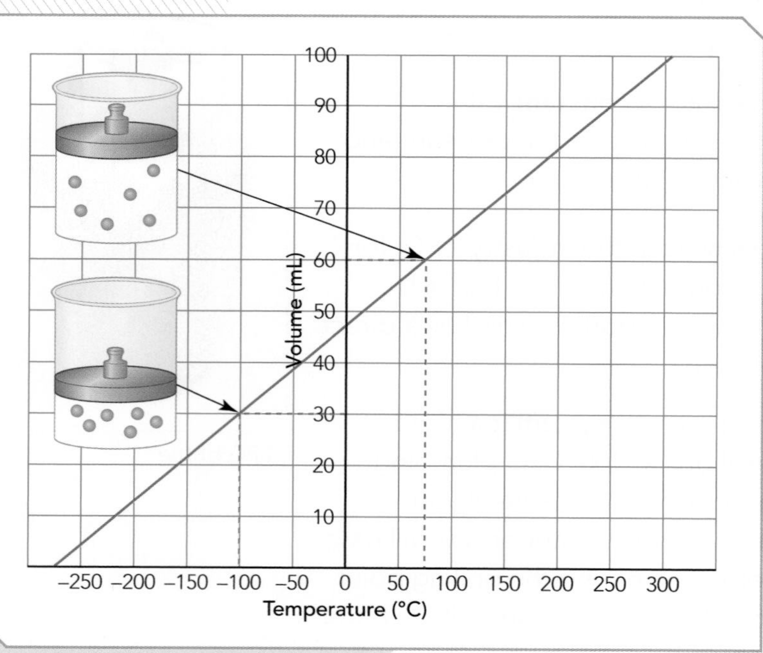

Graph showing Charles's law

Work in groups of three to answer the following questions.

1. Explain how to make petroleum gas into a liquid using Charles's law.

2. Why does the top of an LPG bottle feel cold?

3. Why are the walls of gas bottles so thick and made of metal rather than plastic?

Properties of heat and how energy is transferred

Specific heat capacity is the amount of heat required to raise 1kg of a given material by 1°C. This will tell us how many kilojoules of energy we use to increase the temperature of the water in the cylinder for instance. You need to understand the basics of this for your Level 3 course, where you will use the information for sizing heat exchangers, etc.

Specific heat capacity

Material	kJ/kg °C
Water	4.186
Aluminium	0.887
Cast iron	0.554
Zinc	0.397
Lead	0.125
Copper	0.385
Mercury	0.125

The above table shows the specific heat capacity values given to common materials associated with plumbing.

If it takes 4.186 kJ to raise the temperature of water by 1 °C:

1. How many kJ would it take to raise the temperature of water by 3 °C?
2. How many kJ would it take to raise the temperature of lead by 7 °C?
3. How many kJ would it take to raise the temperature of copper by 15 °C?

Show your workings.

Expansion and contraction of materials

All the materials you work with within the plumbing industry will expand and contract in response to temperature change. Have you heard the sound of the expansion of heating pipes that have not been correctly installed, or plastic guttering creaking in the sunlight? The simple reason for this is that when they are heated the molecules of the materials move further apart, taking up more room. When material contracts the reverse happens. The molecules move back closer together. If we don't take into account this process when installing a system, it could lead to problems, such as leaks. All materials have a **coefficient** value, which we can use to help calculate how much they will expand in response to change in temperature.

key terms

Coefficient: a physical property that is constant under specified conditions.

Coefficient value

Material	Coefficient °C
Plastic	0.00018
Zinc	0.000029
Lead	0.000029
Aluminium	0.000026
Tin	0.000021
Copper	0.000016
Cast iron	0.000011
Mild steel	0.000011
Invar	0.0000009

length (m) × temperature rise (°C) × coefficient value of material = the total the material will expand

Example: 10 m of copper × 30 °C rise in temperature × 0.000016 (Value) = 0.0048 m or 4.8 mm

Remember: 1 m = 100 cm = 1,000 mm

ACTIVITY

What is the overall **thermal** expansion of plastic if you have a temperature rise of 15 °C? Show your workings.

key terms

Thermal: relating to or caused by heat.

Heat transfer

There are three methods of heat transfer: *conduction*, *convection* and *radiation*.

ACTIVITY

In the table below, define the processes of conduction, convection and radiation and give an example of where their effects are seen in mechanical services systems.

Method of heat transfer	Definition of process	Examples of systems in the mechanical services
Conduction		
Convection		
Radiation		

Force and pressure

Pressure is measured in newtons per square metre (N/m^2) or pascals (Pa). One pascal equals 1 N/m^2. You will also hear of other ways to express pressure, such as 'bar' and 'pounds per square inch' (PSI – lb/in^2). These are generally used when we are testing systems, such as heating pressures or unvented hot water systems. Cold water mains are also measured in bars, and PSI is used for expansion vessels.

1 bar = 100,000 N/m^2

1 lb/in^2 = 6,894 N/m^2

1 Pa = 1 N/m^2

Atmospheric pressure is measured at sea level and equals 101,325 N/m^2 or roughly 1 bar.

Pressure is directional. A solid material, such as a lump of lead, for instance, will create a downward force directed to the ground. Liquids, however, will exert a pressure all the way around their container/vessel sides and base. Gases exert pressure in all directions. Understanding the pressure of the system you are working with will allow you to fit the correct components.

ACTIVITY

1. A system is set at 24 lb/in^2. What is the value in newtons per square metre?

2. Express 300,000 N/m^2 in bar.

There is also something called 'head of pressure'. Have a look at this diagram:

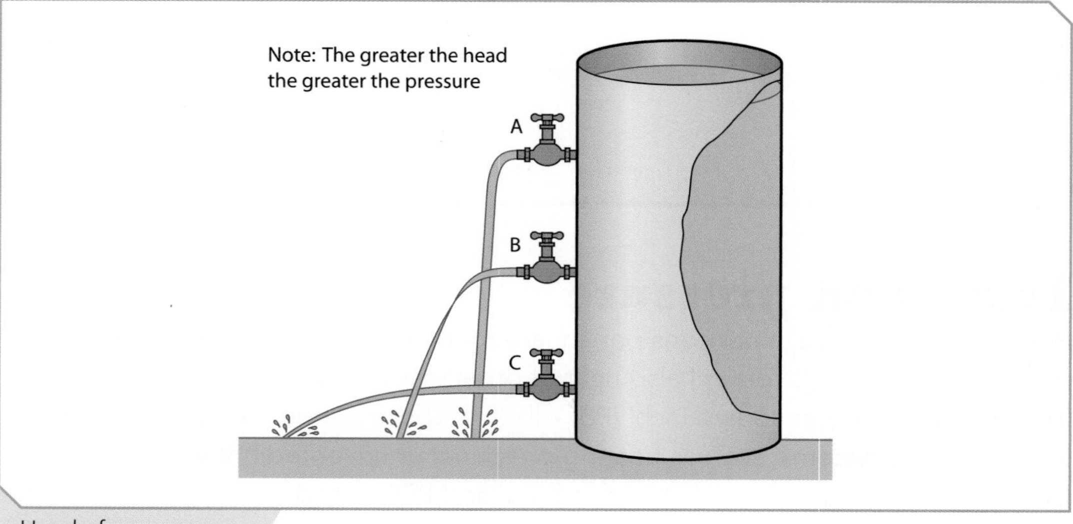

Head of pressure

ACTIVITY

From the information in the previous diagram state in the table below what is happening to the water coming out of the tap.

Tap	What is happening to the water from the tap in relation to pressure?
A	
B	
C	

You can use a 'rule of thumb' to estimate head of pressure. This rule is taken from a measurement in metres of water stored at a height to where the water will **terminate**. This measurement can then be converted into bars.

For instance, 10 m 'head of pressure' would have roughly 1 bar of pressure.

key terms

Terminate: to come to an end of a process. For instance when running water through pipes out to a tap, the tap would be the termination point.

ACTIVITY

Look at the diagram below. Which has the highest pressure at the tap? Circle your answer and give an explanation why. Use the terms 'volume' and 'head' in your answer.

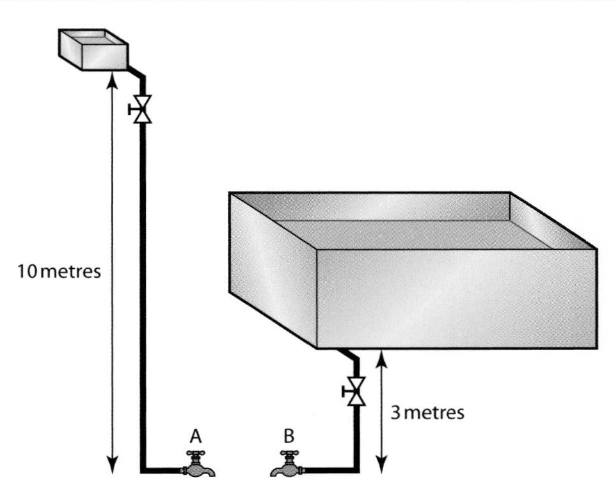

Two cistern pressure

Tap A

Tap B

Explanation:

ACTIVITY

Using the diagram on the previous page again:

Tap A has 15 mm pipework

Tap B has 42 mm pipework

Explain why the size of the pipework will affect the flow rate (l/s) to the tap? Use the terms 'velocity' and 'flow rate' in your answer.

Explanation:

ACTIVITY

In small groups, discuss what else can be used to increase the pressure of water in a system. Make a note of your conclusions.

Other factors need to be taken into effect when we talk about velocity and **flow rates**.

> **key terms**
>
> **Flow rate:** the volume of flow per unit of time. We use litres per minute in the mechanical sector.
>
> **Siphonage:** a process using a U-bend to draw fluid from one container to another at a lower level by gravity.

ACTIVITY

Outline the effects the following will have on the velocity and flow rates:

Reason	Effect this will have on the flow and velocity
Using fittings instead of machine bends	
Change of direction of pipework	
Pipe size increase	
Burred pipe after cutting	
Reducing pipe size	
Introducing valves that are not full bore	

Siphonage and capillary action

Siphonage

Moving water from one vessel to another without the use of a tap can only be achieved by using a siphon. We see this every day when we use siphonic WCs. Atmospheric pressure on a liquid will be constant, as in a cistern in a WC.

Siphonic action toilet

ACTIVITY

Explain the siphonic process by showing that you understand the workings of a siphon.

key terms

Capillary action: the tendency of a liquid in a small space or absorbent material to rise or fall as a result of surface tension.

Capillary action

Capillary action is an important factor in the movement of water. It can take place naturally or be created by the introduction of heat. When you use end feed fittings, notice where the solder is going. You will see that it gets sucked into the small gap between the two materials. Water also does this because it wets surfaces (adhesion), and gets pulled through small gaps by surface tension. Heating can increase the rate of capillary action.

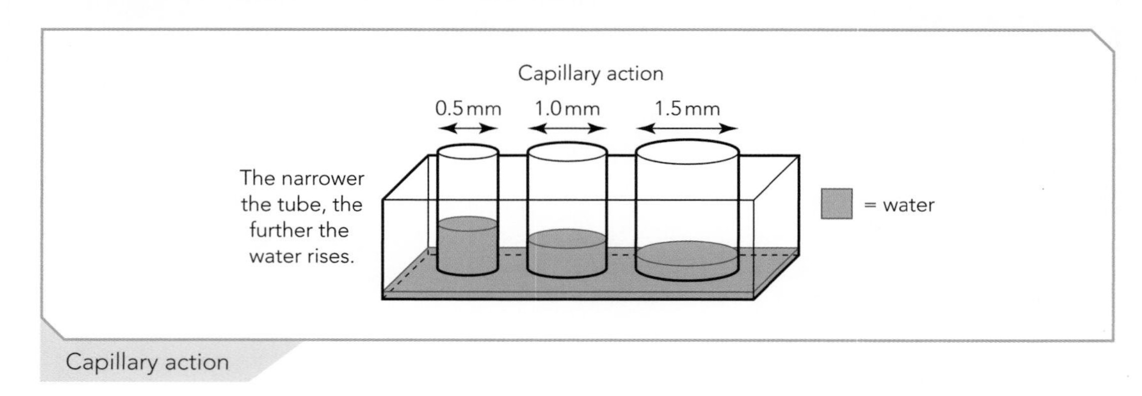

Capillary action

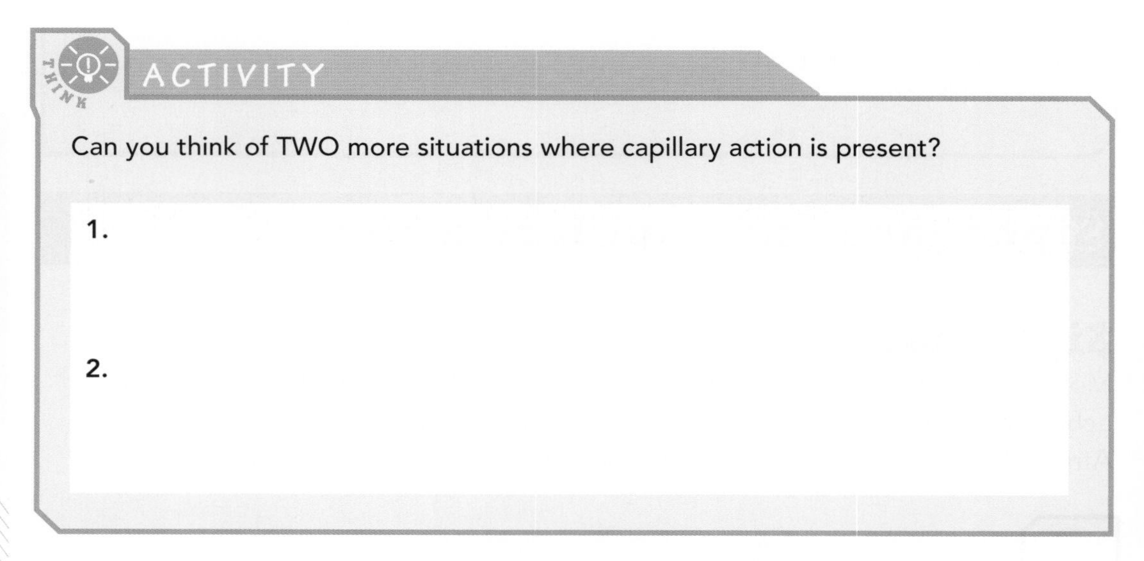

ACTIVITY

Can you think of TWO more situations where capillary action is present?

1.

2.

Materials and their properties

There are three types of force that materials can withstand: *compression*, *tension* and *shear*. Compression is a pushing force such as that experienced by the supports under a bridge. Tension is a stretching force, for instance as occurs in a crane cable when lifting heavy objects. Shear force is a force that is applied at right angles to a surface, for instance when materials are torn. All materials have a limit as to how much *stress* (force per unit area) they can withstand. This is important because you will be using various materials for pipework and you will need to know their limits. We measure stress in Newtons per square metre, N/m^2. Using the correct material to withstand certain stresses will prevent your work being damaged or creating damage. Understanding the terms given to the materials regarding their strength or their elasticity will speed up your decision making.

ACTIVITY

In small groups, discuss what materials are used for pipework in the plumbing industry and why they are suitable for the task.

ACTIVITY

Make a short statement about the properties of the following materials. You will need to find out the definitions of the words below, using the internet to help. You will need to use at least one of these definitions in your statement for each pipework type:

elasticity, plasticity, durability, ductility, malleability, tenacity, thermal expansion

- Plastic polythene heating tubes
- Copper
- LCS (low-carbon steel)
- Lead

key terms

BS EN: British Standard and European number. British Standards ensure the standards of quality and the dimensions are constant and correct. Where an EN has been input, this means the standard has been met in Europe.

Metals that are used by the mechanical sector

Copper

Copper is the most commonly used material for domestic plumbing. Copper piping can come in a vast range of sizes from 6 mm diameter upwards. The most common sizes for everyday use are 22 mm and 15 mm tubing. Copper tubing comes in four different types, which you will investigate in the next activity. Copper tubing has a **BS EN** 1057 number to ensure you know it meets the standard required.

The four types of copper

Fill in the table below using the webpage shown and others to describe the FOUR types of tube produced from copper.

http://www.ukcopperboard.co.uk/

We have filled in the details for the first type for you.

Table Types	Specific **R number**	Hard/soft/both	What it is used for
W	*R220*	*Soft*	*Micro-bore heating systems*

key terms

R numbers: measure the temper of material.

Bore: internal hole in a tube of a given size for something to travel in.

Low-carbon steel, galvanised pipes and other materials

Low-carbon steel, which is also known as LCS or barrel pipe, is used most in industrial premises. It has a BS 1387 number attached to it and comes in many different weights dependent on the purpose to which it will be put. It is also colour coded to match the grade, but you must check carefully as the tubes now come painted with red oxide. There will be more on this later.

The three grades of LCS

Fill in the table below using the internet and other literature to help you. We have done one for you.

Grade	Colour code	Usage	Wall thickness/bore
Light LCS			
Medium			
Heavy	*Red*	*Gas*	*Thick walls/small bore*

Other materials found in the industry

Answer the following questions, using as many sources of information as you need.

Galvanised tubing

1. Why doesn't galvanised tube rust?
2. What is an alloy?
3. Name TWO different ways to connect two different lengths of galvanised tube.

Lead

4. Why is lead still important to plumbers?
5. Why isn't lead used any more for pipework?
6. What is meant by the term 'malleable'?

Cast iron

7. Where will you find cast iron pipework?

8. What TWO materials make up cast iron alloy?

9. What is meant by the term 'brittle'?

Plastics

10. What are the TWO main categories of plastic used in the industry?

11. Where do we use plastics in plumbing?

12. What is meant by the term 'UV damage'?

Alloys

Alloys are a combination of two or more metals. The different combinations of metal alloys give them different properties for application in different situations. For instance, the alloy used in car wheels or rims is made of aluminium and magnesium. This alloy is also used for aircraft production. It is very strong and light and doesn't rust.

In the mechanical sector, we also use a vast array of alloys, perhaps without being aware of it.

Identifying the correct alloy

Complete the table below, indicating what the alloy is made of and where it is used:

Name of alloy	Made of	Main use(s)
Gunmetal		
Solder		
Brass		
Bronze		

Corrosion of materials in the mechanical sector

It is vital to remember that corrosion can affect various types of materials although metals are more at risk than other materials, such as plastic. Water and air will start the corrosion of **ferrous** metals and will result in what we know as rust. Low-carbon steel uses a covering of red oxide to deter the effects of normal atmospheric corrosion. Another reason for corrosion is the presence of acids in the water. Non-ferrous metals, such as copper and lead, do not contain iron and therefore will not rust. They will however create a barrier called a **patina**; this barrier gives the metal a dull colour compared to the shiny surface of new materials. It's important to clean and remove this barrier before soldering or welding where needed.

key terms

Ferrous: containing iron.

Patina: a film of oxide that forms on some metals.

ACTIVITY

Forms of corrosion in plumbing

1. In small groups, discuss why a central heating system can create hydrogen gas. Use the internet to carry out some research before your discussion.

2. Explain what ferrous oxide is.

3. Where there is lead pipework still in a dwelling and the water type is acidic, what is happening to the lead?

4. Why is it important to replace old lead piping?

key terms

Dissimilar: not alike; not similar.

Electromotive series: metals ranked in order of their ability to corrode other metals by a battery effect.

Anode: positively charged electrode.

Cathode: negatively charged electrode.

Electrolytic action

Electrolytic action is created when two **dissimilar** metals in the **electromotive series** create a battery effect. It corrodes the metal that is lower in the series. A typical example of this would be the use of a galvanised CWSC and copper. The diagram below shows how the electrically charged ions flow from an **anode** (+) to a **cathode** (–) through an intermediate source known as the electrolyte, which for the plumbing industry is usually water.

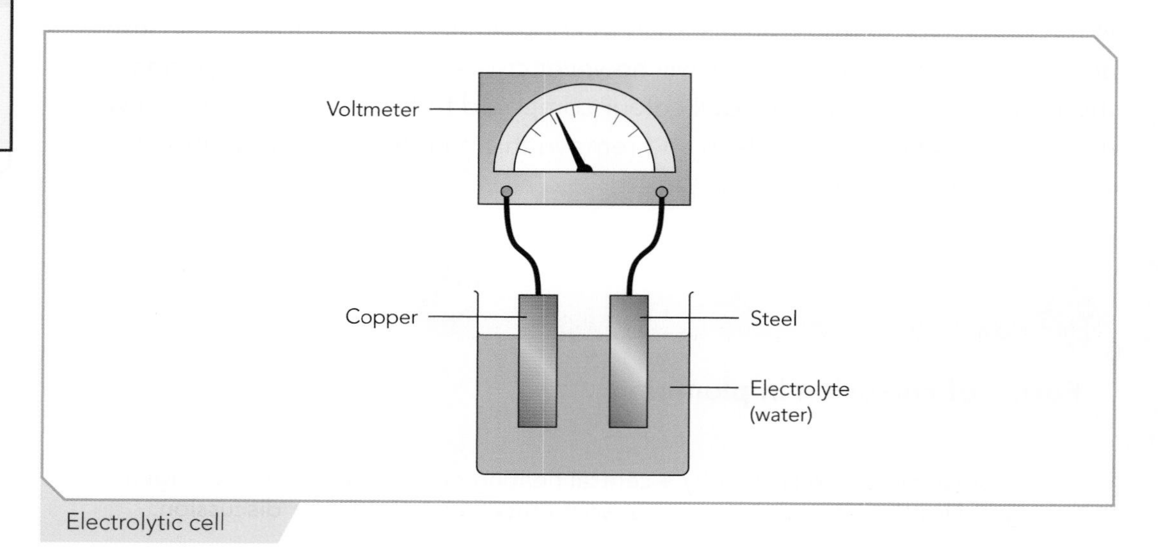

Electrolytic cell

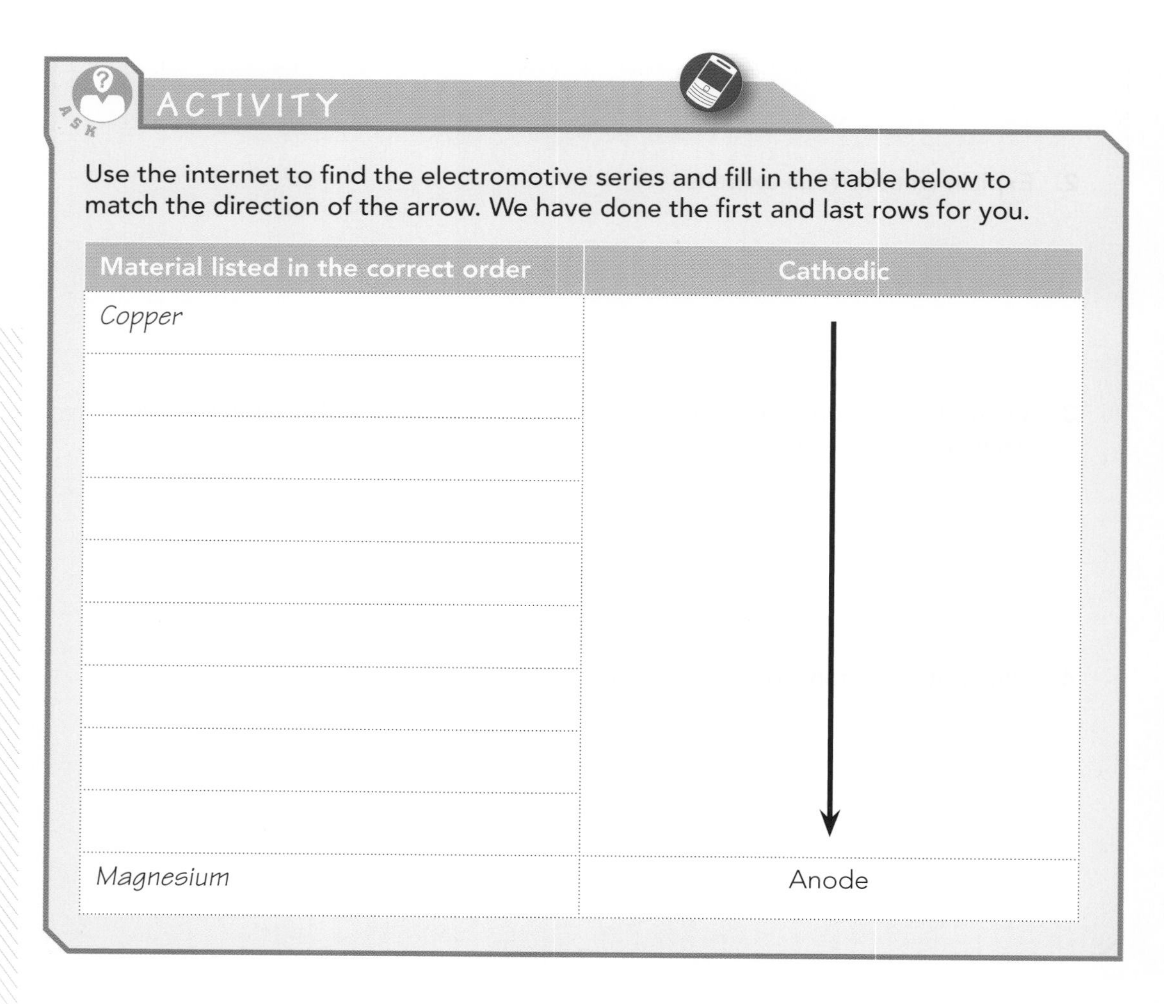

ACTIVITY

Use the internet to find the electromotive series and fill in the table below to match the direction of the arrow. We have done the first and last rows for you.

Material listed in the correct order	Cathodic
Copper	
Magnesium	Anode

Basic principles of electricity

You should know that water and **electricity** do not mix. However, there are many occasions in the plumbing industry when they come into close contact. There are many rules for working with electrics and in this section we will be going through the basic principles of electricity and looking at the relationship between the two sectors.

Below is a list of SI units that are relevant to electricity:

Electricity SI units

SI unit	Measure of	Symbol
coulomb	charge	C
joule	energy	J
ohm	resistance	R
volt	potential difference	V
watt	power	W
ampere	electric current	A

key terms

Electricity: the flow of electrons through a conductor.

ACTIVITY

Use the internet to research what a conductor is. Find out what makes a material a good conductor.

Electrical energy

The basic unit of work or energy is the joule (J), but this is a very small unit and only represents a power of 1 watt operating for 1 second. To avoid silly calculations (imagine if we measured our working day in seconds, not hours), we therefore use a much larger unit.

The unit used for electrical energy is the kilowatt-hour, which represents 1 kilowatt for 1 hour. From this we can see that:

1 joule (J) = 1 watt (W) for 1 second (s)

1,000 J = 1 kilowatt (kW) for 1 second

In 1 hour there are 3,600 seconds. Therefore:

3,600 s × 1,000 J = 1 kW for 1 hour (kWh)

The kilowatt-hour is the unit used by the electrical supply companies to charge their customers for their use of electrical energy supplied.

ACTIVITY

When you get home tonight after college have a look around your house. Check the distribution unit and the electricity meter.

You'll see that the electric meter is measuring in kWh. However, these are more often referred to as 'units' by the time they appear on your bill!

Measuring electricity

When working with electricity you need to understand the terms used, such as *current*, *voltage*, *resistance*, *amperage*, *ohm* and *voltmeter*. All of these words are related to the terms we use to measure electricity.

ACTIVITY

Can you match each term listed below with its correct definition by drawing lines from one to the other?

Term	Definition
Resistance	The resistance to current flow in a circuit is measured by this unit and is given a Greek symbol Ω
Voltage	The quantity of electricity that flows every second, which is measured in amps and given the symbol *I*
Ohms	If a conductor is large, the current has less difficulty moving on. The smaller the conductor the greater this factor becomes. Normally given the symbol *R*
Current	The unit for measuring electric current. Given the symbol A
Amperes	This measurement is to identify the potential difference of energy that is used by the circuit. Given the symbol *V*

The equation that describes the relationship between potential difference (voltage), current (amperage) and resistance is:

$$V = I \times R$$

This is known as Ohm's law.

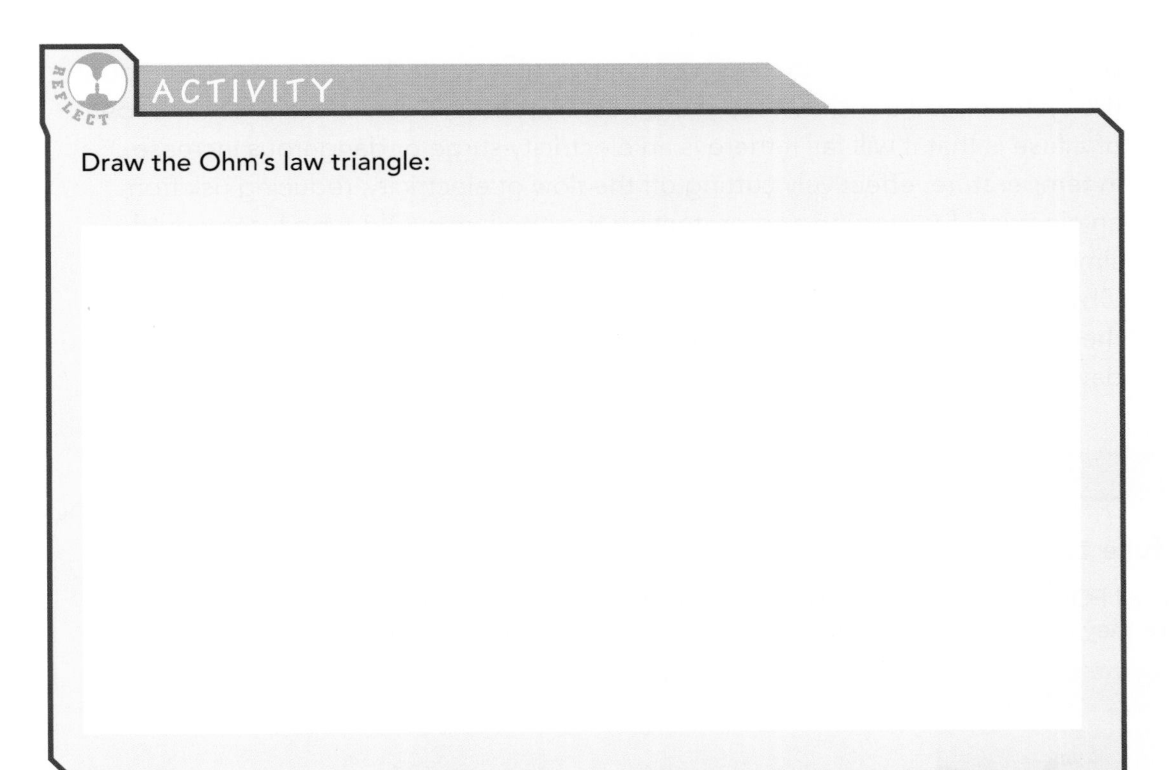

Draw the Ohm's law triangle:

How to use the Ohm's law triangle

To use the triangle you will need to know two of the quantities to work out the third. Let's say we know what the voltage (V) and the resistance (R) are, but we need to find the amperage, or current (I).

We would cover the I and the R symbols on the triangle to check we are going to get the correct unit we need.

The equation for this would be $I = V/R$

ACTIVITY

Complete the equations below:

1. $V = I \times R = 2 \times 120 =$ Voltage
2. $I = V/R = 300/100 =$ Current
3. $R = V/I = 240/120 =$ Resistance

Watts and fuses

Another unit we need to know about is the watt, which measures power. This is linked to the rating of fuses. All electrical appliances that can be plugged into the mains, such as televisions, computers, fridges, etc. are power rated using watts to show the power that is consumed by that appliance. They will also have a fuse fitted – this is to protect the wiring from surges of high electrical current passing through. When there is a constant high electrical current passing through the system it will transmit heat. If this heat increases, it can cause a fire. The idea of a fuse is that it will fail if there is an electricity surge or dangerous increase in temperature, effectively cutting off the flow of electricity, reducing risk from an electrical fire starting and protecting the appliance. Old-type fuses would simply be a piece of wire calculated to burn away if the temperature was raised. Obviously this could be altered by putting in a thicker piece of wire. This would therefore risk the outbreak of an electrical fire or damage to the ring circuit. These days, fuses are a bit more sophisticated.

ACTIVITY

Identifying the fuse type

You will come across all FOUR main types of 'fuse' in your work. Complete the table below indicating how they work and where they are most likely to be used. State why a circuit breaker is different from a fuse.

Type	How it works and where it is used
Cartridge	
Rewireable fuse	
Portable residual circuit breaker (RCB)	
Miniature circuit breaker (MCB)	

All fuses are rated in amps. To work out the correct fuse size for an appliance you need to use the following formula:

amps = watts/volts

All appliances will have the amount of electricity they consume measured in watts. The current flow is measured in amps. For instance, the equation for a washing machine with a wattage of 150 W on a domestic voltage of 240 V would be:

150 W/240 V = 0.625 amps

So a fuse rating of 0.625 A would be needed.

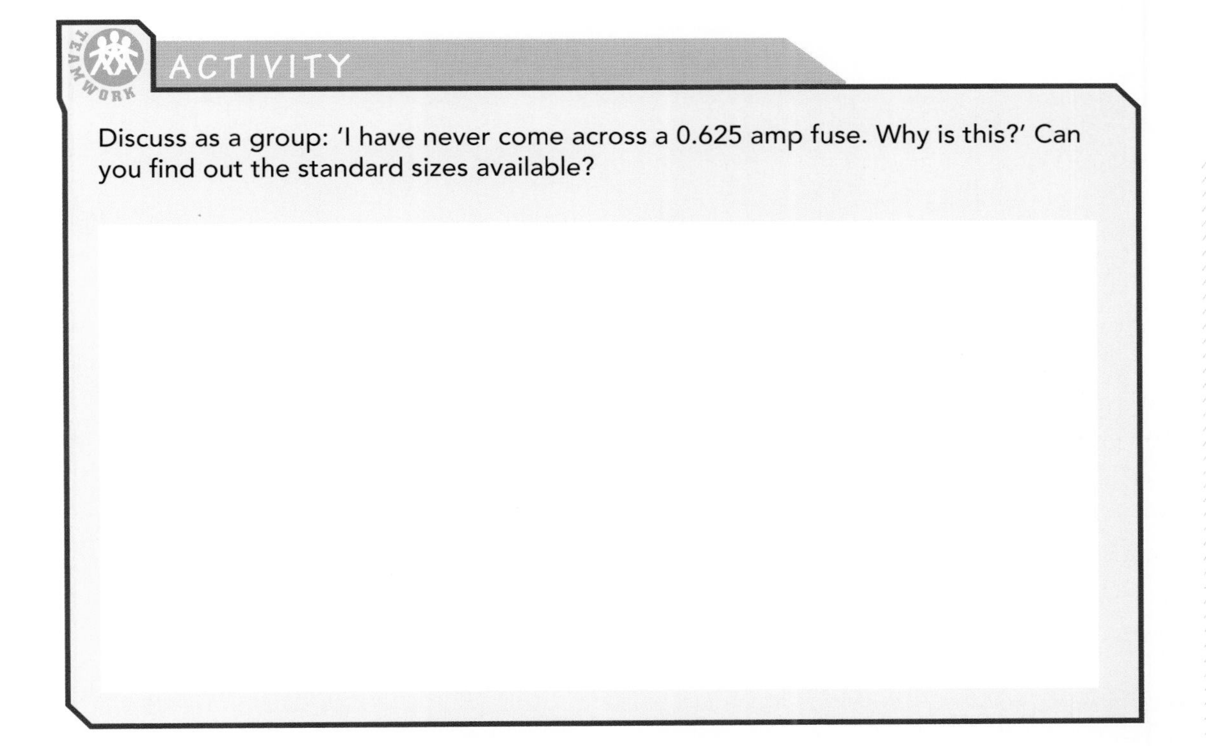

TEAMWORK
ACTIVITY
Discuss as a group: 'I have never come across a 0.625 amp fuse. Why is this?' Can you find out the standard sizes available?

AC and DC

There are two different types of current available to pass on electrons. These are known as *alternating current* and *direct current* or AC and DC.

Direct current: in the diagram opposite the current flows in the same direction constantly. We get this one direction flow typically from a battery, which has an anode and a cathode that when connected create the circuit.

Alternating current: in AC the current moves back and forth. This type of current is produced by an alternator in power stations, whether they are thermal (such as gas-fired power stations) or kinetic (such as wind farms).

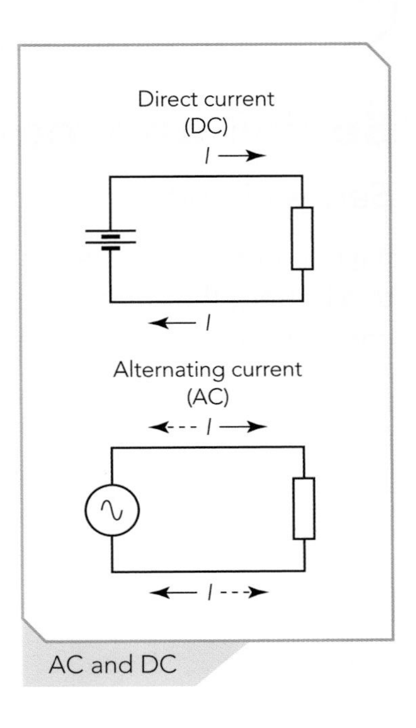

AC and DC

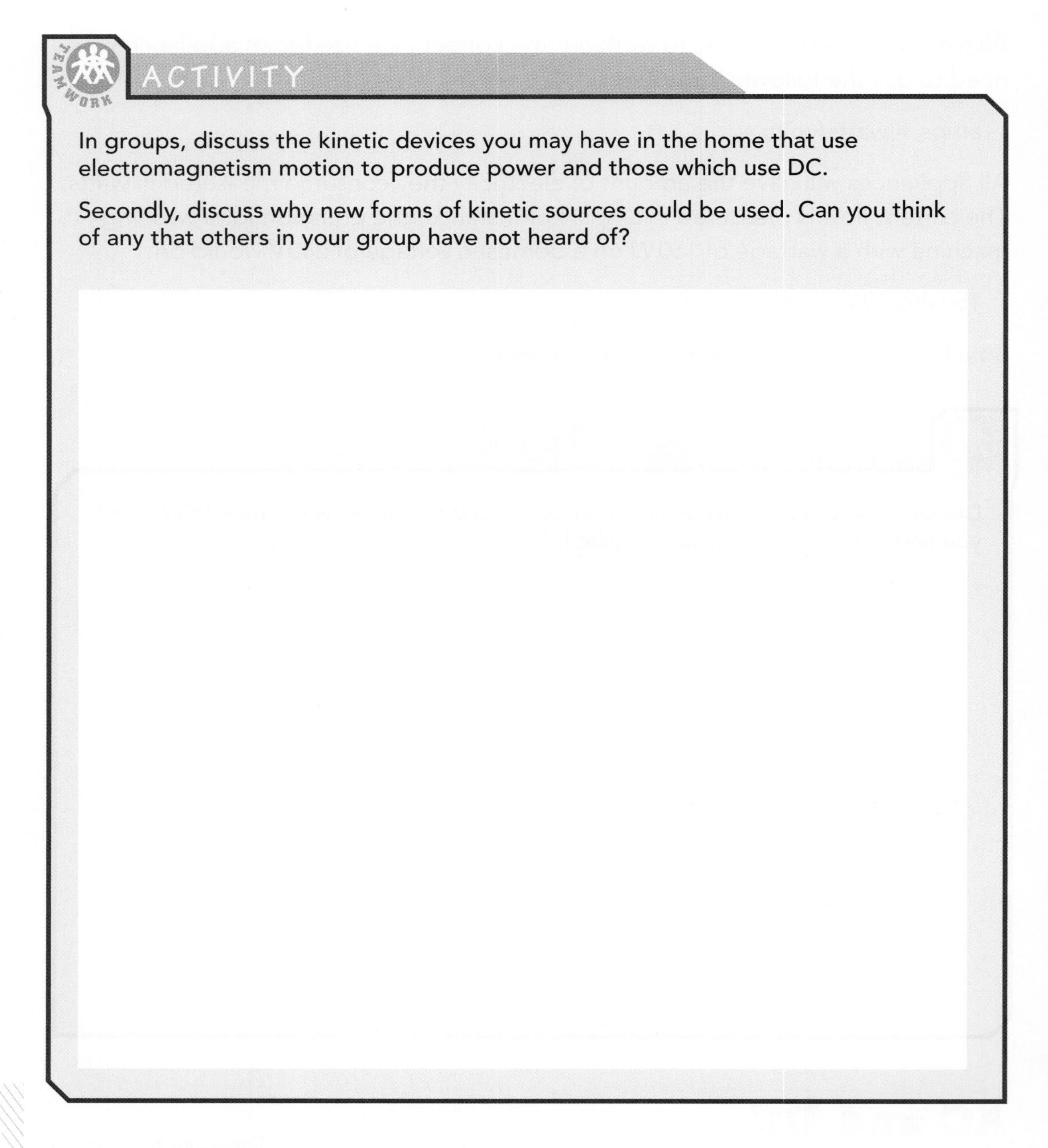

TEAMWORK
ACTIVITY
In groups, discuss the kinetic devices you may have in the home that use electromagnetism motion to produce power and those which use DC.
Secondly, discuss why new forms of kinetic sources could be used. Can you think of any that others in your group have not heard of?

Series and parallel DC circuits

Series circuits

If a number of resistors are connected together end to end and then connected to a battery, the current can only take one route through the circuit. This type of connection is called a series circuit as shown below:

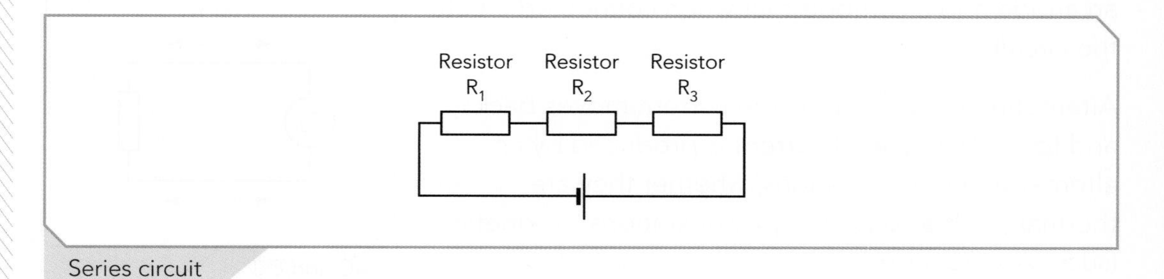

Series circuit

The following rules apply to a series circuit:

The total circuit resistance (R_t) is equal to the sum of all the circuit resistors. In other words, to find the total resistance of the circuit, we add up the value of the individual resistors. In our diagram, this would be:

$$R_t = R_1 + R_2 + R_3$$

The total circuit current (I) is equal to the battery power divided by the total resistance.

This is just Ohm's law again, which if you remember is $I = V/R$.

The current will have the same value at any point in the circuit.

The potential difference across each resistor is proportional to its resistance.

If we think back to Ohm's law, we use voltage to push the electrons through a resistor. How much we use depends upon the size of the resistor.

The bigger the resistor, the more we use. Therefore:

$$V_1 = I \times R_1,\ V_2 = I \times R_2, \text{ etc.}$$

The supply voltage (V) will be equal to the sum of the potential differences (pd) across each resistor. In other words, if we add up the pd across each resistor (the amount of volts 'dropped' across each resistor), it should come to the value of the supply voltage. We show this as:

$$V = V_1 + V_2 + V_3$$

The total power in a series circuit is equal to the sum of the individual powers used by each resistor. In other words:

$$P = P_1 + P_2 + P_3$$

Parallel circuits

If a number of resistors are connected together as shown below, they are said to be connected in parallel.

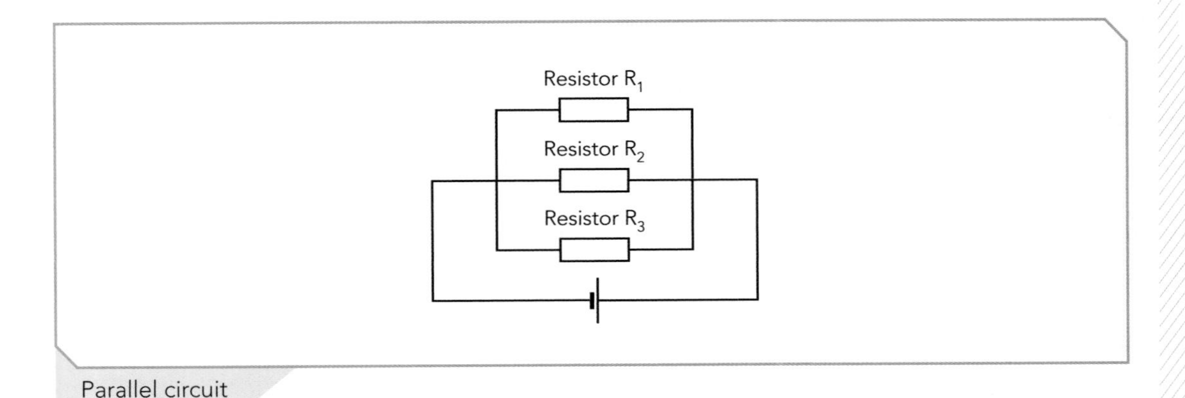

Parallel circuit

In this type of connection, the total current splits and divides itself among the different branches of the circuit. However, it should be noted that the pressure pushing the electrons along (voltage) will be the same through each of the branches. Therefore, any branch of a parallel circuit can be disconnected without affecting the other remaining branches.

In summary, we can therefore say that the following rules will apply to a parallel circuit:

The total circuit current (I) is found by adding together the current through each of the branches: $I = I_1 + I_2 + I_3$

The same potential difference will occur across each branch of the circuit: $V = V_1 = V_2 = V_3$

Where resistors are connected in parallel and, for the purpose of calculation, it is easier if the group of resistors is replaced by one large resistor (R_t): $\frac{1}{R_t} = \frac{1}{R_1} + \frac{1}{R_2} + \frac{1}{R_3}$

ACTIVITY

Answer the following questions related to series and parallel circuits.

1. If an electric fire of resistance 24.8 Ω, an immersion heater of resistance 34.8 Ω, a small microwave oven of resistance 45.9 Ω and a toaster of resistance 120 Ω are connected to a 230 V power circuit, calculate the current taken by each appliance and the total current drawn from the supply.

2. A 230 V electric kettle has a resistance of 88 Ω and is connected to a socket outlet by a twin cable, each conductor of which has a resistance of 0.1 Ω. If the total resistance of the cable from fuse board to the socket is 0.8 Ω, calculate the total resistance of the whole circuit.

3. Calculate the resistance and the current drawn from the supply by the following equipment connected to a 230 V supply:

   - a 4 kW 230 V immersion heater

   - a 600 W 230 V microwave oven

   - a 1 kW 230 V electric fire

   - a 750 W 230 V stereo system

Earth continuity

The fundamental rule for ensuring that any system is **earthed** is to ensure the safety of whoever may come into contact with it. Anything that is fitted directly to the main system will have an earth wire; plugs in an appliance will have an earth wire.

The main systems TT, TN-S and TN-C-S will be fitted by a fully qualified electrician.

key terms

Earthing: transferring of any leak of electricity or charge direct to the ground for safety.

ACTIVITY

When you get home tonight after college have a look around your house. Check the distribution unit and the electricity meter.

You'll see that the electric meter has an earth cable; please do not touch anything that you are unsure of or take things apart. If you can't find any signs there, go to your gas meter. If there isn't an earth wire there, then there should be!

If you are working with an oil system, this should also be earthed; you may be lucky enough to see the earth rod as well.

As a mechanical engineer you will come across 'supplementary bonding', where clips or clamps are attached to all exposed metalwork, which will also be relaid to earth.

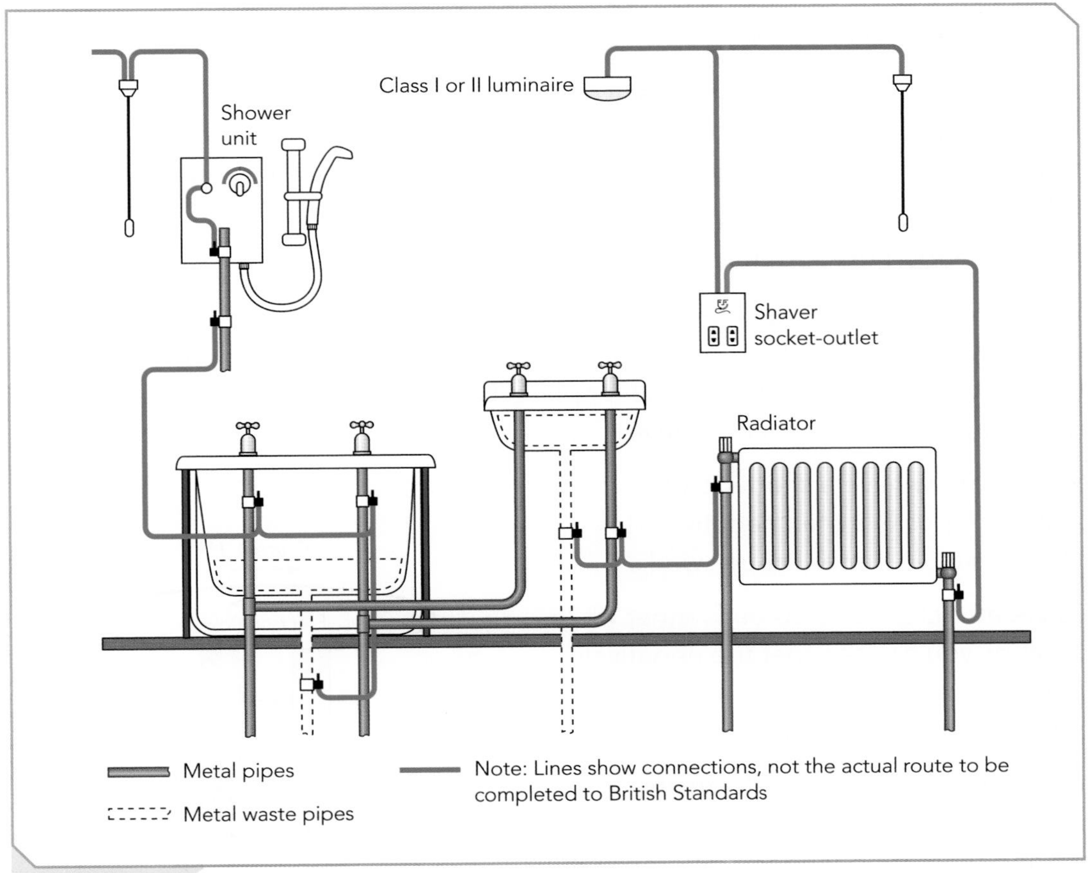

Earthing

You have been asked to remove a piece of pipework that is known to be bonded. In pairs, discuss what you should do to avoid breaking the 'bond' and putting yourself in danger.

Your questions answered...

I have often heard that current does not flow positive to negative but the other way around. Can you please explain this to me?

The convention is to say that the current flows from positive to negative, but this was decided on before it was known that the electric charge is carried around the circuit by tiny particles called electrons. These are negatively charged, so they in fact flow around the circuit in the opposite direction to the conventional flow of current shown on circuit diagrams.

We now know that a force is needed to cause this flow of electrons.

This has the quantity unit symbol V (volt). If we take two dissimilar metal plates and place them in a chemical solution (known as an electrolyte), a reaction will take place in which electrons from one plate travel through the electrolyte and collect on the other plate. Now, one plate has an excess of electrons, which will make it more negative than positive.

Of course, the other plate will now have a shortage of electrons, which makes it more positive than negative. This process is the basis of how a simple battery or cell works.

Because unlike charges are attracted towards each other while like charges repel each other, you can see that the negatively charged electrons will be driven away from the negative plate, through the conductor towards the positive plate.

This drift of free electrons is what we know as electricity and this process will continue until the chemical action of the battery is exhausted and there is no longer a difference between the plates.

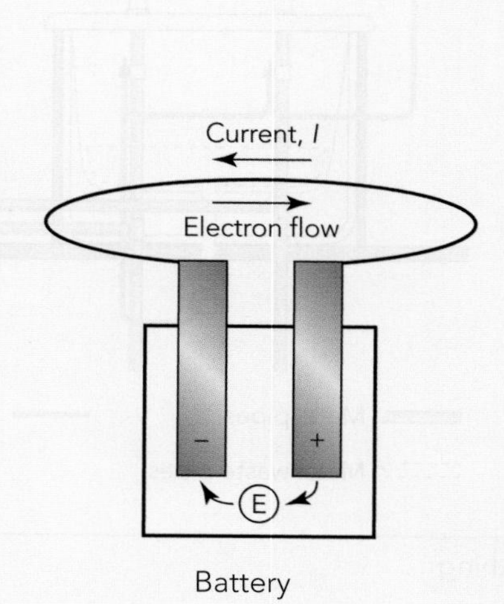

QUICK QUIZ

1. What do the letters H and O stand for in the equation H_2O?
2. What are the SI units for mass, weight and pressure?
3. What is the pH value of distilled water?
4. What is meant by 'capacity' in terms of volume?
5. What is the name of the surface that is created on lead and copper for protection?
6. Lead is a malleable material; what does the word 'malleable' mean?
7. Is copper known as an anode or a cathode?
8. What is measured in degrees Kelvin?
9. How would you work out how dense something is?
10. What is meant by the term 'circumference'?
11. What does kJ stand for?
12. If you had a length of pipe 2,500 mm long, how many centimetres would it measure?
13. What does RCB stand for?
14. What is temporary bonding and why is it used?

Unit 005/205

Understand and carry out site preparation, and pipework fabrication techniques for domestic plumbing and heating systems

It is important to know the various ways to connect, run and support all the different types of pipework available. Understanding where and how to fit pipework is important – poorly made fittings can lead to leaks and unsecured pipework can lead to stress on the system. Choosing the correct tools to assist you will speed up the process of fitting, therefore making the job easier.

Before pipe bending machines were invented plumbers would use a spring to help them bend pipework and, although it is not the best method, it is still used by some people and can be useful in restricted spaces where you can't use a stand bender. Your tutor will show you the various methods of using hand benders. Although some people can see how to use them straight away, for some it may take longer to 'click', so be patient. We will not go through how to use a bender in this unit, but will mention the principles, along with other methods used for different materials.

Learning outcomes

- Know the types of hand and power tools used for domestic plumbing and heating work
- Know the types of domestic plumbing and heating pipework and their jointing principles
- Know the general site preparation techniques for plumbing and heating work
- Be able to apply general site preparation techniques for domestic plumbing and heating work
- Know how to use clips and brackets to support domestic plumbing and heating pipework and components
- Be able to apply fixings and brackets to domestic plumbing and heating pipework and components
- Know the installation requirements of domestic plumbing and heating pipework
- Be able to install domestic plumbing and heating pipework
- Know the inspection and soundness testing requirements of domestic plumbing and heating pipework
- Be able to inspect and soundness test domestic plumbing and heating pipework

Key knowledge

- Plumbing tools
- Pipework bending and jointing processes
- Health and safety in the workplace
- Fixings, supports and brackets
- Preparing building fabrics
- Symbols used in plumbing
- Soundness testing pipework before commissioning

key terms

PTFE: polytetrafluoroethylene is a thin plastic which comes on a roll, and is used in the jointing process.

Tight: term sometimes used to describe a fitting that has been made correctly.

Pipework jointing techniques

ACTIVITY

The table below lists all the different types of pipework you will have to joint together at some time. Select the most appropriate method by ticking the relevant column, and then justify your choice. If there is more than one correct method, write 1, 2, 3 or 4 in the appropriate columns in order of preference. State the type of system the pipework can be used with. If there are limitations for using the pipe and fitting, state this too.

We have done the first one for you.

	Solder joint	Crimped joint	Manipulative joint	Compression joint	Pushfit joint	Threaded joints	Welded joints	Solvent weld	Mechanical joint	Justify your choice Limitations of use System most suited to tube type
LCS				✓ 2		✓ 1				*Threading the pipe and using hemp and paste or* **PTFE** *is the most common and secure method of ensuring that the fitting is* **'tight'**. *Compression fittings could become loose through expansion and contraction. Not to be used for drinking water. Best used for industrial heating systems.*
Grade X copper										
Grade Y										
Grade Z										

	Solder joint	Crimped joint	Manipulative joint	Compression joint	Pushfit joint	Threaded joints	Welded joints	Solvent weld	Mechanical joint	Justify your choice Limitations of use System most suited to tube type
Grade W										
Polythene										
ABS										
Polypropylene										
UPVC										
Polyethylene										
Lead										
Steel pipe										

Plumbing jointing processes

ACTIVITY

In the spaces provided below, draw a diagram or paste in a photo of a plumbing fitting related to the jointing type. In the final column, state the tools you would need to make the joint – from cutting the pipe to finishing the joint.

Jointing type	Diagram/picture	Choice of tools to make the fitting, starting from cutting the pipe to a joint being made
Soldered (**end feed**)		
Pushfit (copper)		
Solvent weld		
Pushfit (plastic waste)		
Threaded		

key terms

End feed: the most commonly used type of capillary fitting used with solder.

Pushfit: fittings used for high-speed installations that can be taken apart again if needed. They use a series of grips and 'O' rings to create the seal.

Threaded: describes the jointing technique for LCS pipework.

key terms

Manipulative: fitted to allow a small amount of movement due to its use or location.

Yorkshire (fitting): manufacturer's name to describe a solder ring fitting.

Crimped: fitting that is compressed on to a pipe using no other substance or specialist parts, such as olives.

Jointing type	Diagram/picture	Choice of tools to make the fitting, starting from cutting the pipe to a joint being made
Compression (copper)		
Compression (LCS)		
Manipulative type B		
Lead		
Steel pipe		
Soldered (**Yorkshire**)		
Crimped		

ACTIVITY

In your workshop, make ONE joint from each of the types listed below. Put the tools and processes you identified into practice. Make THREE attempts and after each attempt record any problem that you faced, or what the outcome was in the table below. Describe how you put right any problems you encountered.

Afterwards, set yourself targets if you need more practice. Remember, practice makes perfect and more than three attempts may be needed before you are comfortable that you are competent with each type.

Joint	Attempt 1	Attempt 2	Attempt 3	Conclusion/ more practice needed
End feed				
Compression type A				
Compression type B				
Yorkshire				
Solvent weld				
Pushfit plastic waste				
Pushfit polythene				
Compression A polythene				
LCS threaded hemp and paste				
LCS threaded PTFE				
Copper crimped				

Tools associated with the installation of BSE materials

Tools

Working in pairs, find the tools that you are likely to come into direct contact with in the workplace hidden in the wordsearch below. If there are any tools you have not heard of, note them down and find what they are and what they are used for.

H	Y	D	R	A	U	L	I	C	B	E	N	D	E	R	O	P	H	A	M	M	E	R	D	S	V	A	M	Y	W	I	I	O	N
A	D	A	D	A	L	A	N	N	A	E	M	M	Y	U	G	T	B	R	A	R	H	D	C	H	E	R	T	L	K	I	P	X	D
C	A	L	C	S	P	I	P	E	C	U	T	T	E	R	K	R	C	G	C	I	S	G	D	R	N	B	C	S	D	E	E	R	M
K	E	C	W	U	N	K	B	O	R	N	F	D	R	E	R	R	I	O	H	N	G	O	O	D	T	F	F	D	P	S	S	V	F
S	D	W	O	O	D	B	O	R	I	N	G	B	I	T	S	E	R	D	I	N	X	V	D	R	K	Y	E	P	H	G	L	T	S
A	F	F	O	L	E	R	T	U	I	O	L	D	D	S	D	N	C	N	N	G	F	R	T	Y	E	Y	R	S	I	R	O	A	P
W	R	G	D	R	D	T	B	Y	T	Y	N	Y	M	U	S	U	U	T	E	R	I	P	S	S	Y	F	U	D	L	F	T	P	R
M	Q	W	C	E	R	C	T	Y	H	G	D	V	H	P	D	P	L	P	B	T	R	R	R	Q	T	R	T	T	I	E	T	E	T
I	I	Y	H	R	R	T	H	T	H	C	V	D	R	S	R	R	A	M	E	N	A	M	E	R	B	R	E	D	P	W	E	M	R
C	R	T	I	E	S	T	E	I	V	E	M	U	D	S	I	E	R	E	N	L	E	R	R	X	E	R	V	D	S	A	D	E	M
H	L	J	S	S	S	E	C	F	S	W	E	W	R	W	L	W	S	D	D	D	F	F	U	Z	C	E	V	B	S	S	S	A	E
E	R	E	E	V	D	D	E	R	R	E	Q	W	E	R	L	N	A	Y	E	Y	Y	L	G	D	D	N	F	R	C	K	C	S	E
A	T	Y	L	T	R	R	R	E	T	S	L	O	B	R	T	E	W	E	R	E	F	R	E	T	T	N	R	E	R	C	R	U	E
L	D	E	D	S	E	E	W	B	T	N	Y	Y	T	Y	N	Y	T	R	E	V	P	L	I	Y	T	A	R	B	E	A	E	R	R
M	E	K	W	W	T	Q	S	C	D	E	F	B	E	R	T	S	S	D	E	D	V	V	F	R	E	P	R	K	W	H	W	E	S
A	A	N	G	G	Y	I	T	Y	N	Y	H	Y	T	E	S	F	F	T	R	R	R	T	P	O	U	S	E	R	D	R	D	D	D
N	H	A	D	F	R	R	K	E	R	F	F	D	F	E	G	H	U	Y	A	U	U	Y	T	R	E	E	E	V	R	O	R	V	R
G	R	M	T	H	Y	R	R	W	R	G	H	F	H	W	G	F	R	T	U	B	N	M	V	C	X	L	X	F	I	I	I	D	D
A	Q	F	W	S	S	F	R	R	A	R	Y	T	U	K	I	R	T	T	Q	R	R	R	V	N	R	B	R	T	V	N	V	R	R
N	E	O	W	E	P	E	F	V	V	S	N	G	R	T	G	D	G	I	S	K	U	Y	N	O	P	A	Y	U	E	U	E	W	L
G	T	E	E	D	I	U	G	U	T	L	E	V	E	L	T	T	T	A	R	D	X	Z	A	W	S	T	S	F	R	J	R	J	L
T	K	L	Q	W	R	E	D	C	R	F	V	L	T	H	N	U	J	M	S	J	U	Y	V	M	N	S	T	H	F	W	D	E	I
H	C	S	J	I	G	S	A	W	W	A	S	C	O	C	S	D	F	C	D	K	S	E	V	B	V	U	X	D	D	X	X	G	R
E	N	I	H	C	A	M	G	N	I	D	A	E	R	H	T	C	I	R	T	C	E	L	E	S	L	J	J	U	Y	T	T	F	D
J	H	Y	T	G	R	F	M	K	I	U	Y	T	R	G	T	N	M	J	K	U	G	Y	N	G	P	D	F	G	T	R	E	H	E
R	K	K	P	L	A	S	T	I	C	P	I	P	E	C	U	T	T	E	R	I	K	O	G	T	Y	A	F	E	W	D	F	H	R
E	H	T	R	B	J	M	I	H	F	E	K	W	E	S	R	C	R	G	G	Y	S	J	U	Y	T	L	D	L	I	U	Y	R	O
E	D	D	K	Q	W	E	R	T	Y	G	K	G	B	F	D	E	E	R	V	L	H	T	F	F	D	D	D	S	V	C	D	E	C
W	G	G	L	L	K	U	I	Y	B	H	L	L	O	P	O	I	U	Y	I	F	E	D	C	S	W	D	F	G	A	J	M	T	D
Y	B	H	A	K	M	L	I	K	E	G	F	D	E	D	E	X	S	T	P	I	P	E	S	L	I	C	E	E	C	W	G	F	N
H	F	O	H	A	N	D	D	I	E	S	L	F	C	V	B	B	S	N	O	R	T	H	A	M	P	T	O	N	T	O	W	C	O
P	E	N	C	I	L	L	B	O	A	T	L	E	V	E	L	P	O	I	U	Y	B	F	C	R	H	G	F	R	T	H	Y	D	M
Y	J	Q	S	C	W	D	V	E	F	B	R	G	N	T	H	M	Y	J	K	U	I	L	O	P	R	T	F	E	E	E	G	T	A
Y	Q	W	E	R	T	Y	U	I	O	P	L	I	K	U	J	Y	H	T	B	A	T	T	E	R	Y	D	R	I	L	L	O	I	I
R	L	K	J	H	G	F	D	S	A	P	O	I	U	Y	T	R	E	W	Q	M	N	B	V	C	X	Z	Z	X	C	V	B	L	D
N	G	L	C	X	A	Q	E	R	E	D	N	E	B	E	R	O	B	I	N	I	M	G	T	H	N	E	B	H	T	T	H	J	J

adjustment spanner
battery drill
boat level
bolster
chalk
circular saw
cold chisel
diamond core drill
electric threading machine
flux brush
gas key
grips
guide
hack saw
hammer
hand dies
hole saw kit
hydraulic bender
Isle of Man key
jig saw
junior hack saw
LCS pipe cutter
level
machine bender
mini bore bender
pad saw
pencil
philips screwdriver
pipe slice
plastic pipe cutter
saw
SDS drill
slotted screwdriver
square
stilsons
tape measure
van
vent key
wood boring bits
wood chisel

Pipework fixing devices

There are various types of fixing devices available to ensure your hard work stays where you and your customer want it. The fixings you use should be able to withstand the most likely causes of strain or damage.

ACTIVITY

Plugs and screws

In groups, discuss the different **plugs** and screws you use on a day-to-day basis, their colours and sizes. Make a list of as many different screw types as you can.

key terms

Plug: device used to aid in the making of a strong fixing, such as a plastic sheath around the screw that fits into drill holes and grips the masonry by means of expansion.

Masonry: stonework, such as brick, concrete and building stone.

Specialist fixings

You will not always make fixings to **masonry** or wood. Sometimes you will be asked to attach an appliance to plasterboard or a poor surface. You will have to choose a suitable fixing type that will take into account the limitations of the surface.

ACTIVITY

Using the pictures of the fixing devices shown below to help you identify them, search the internet to find out where these fixings are used and how they work.

Fixing device	Explanation of where the fixing is used and how it works
Rawl bolt	
Coach screw	

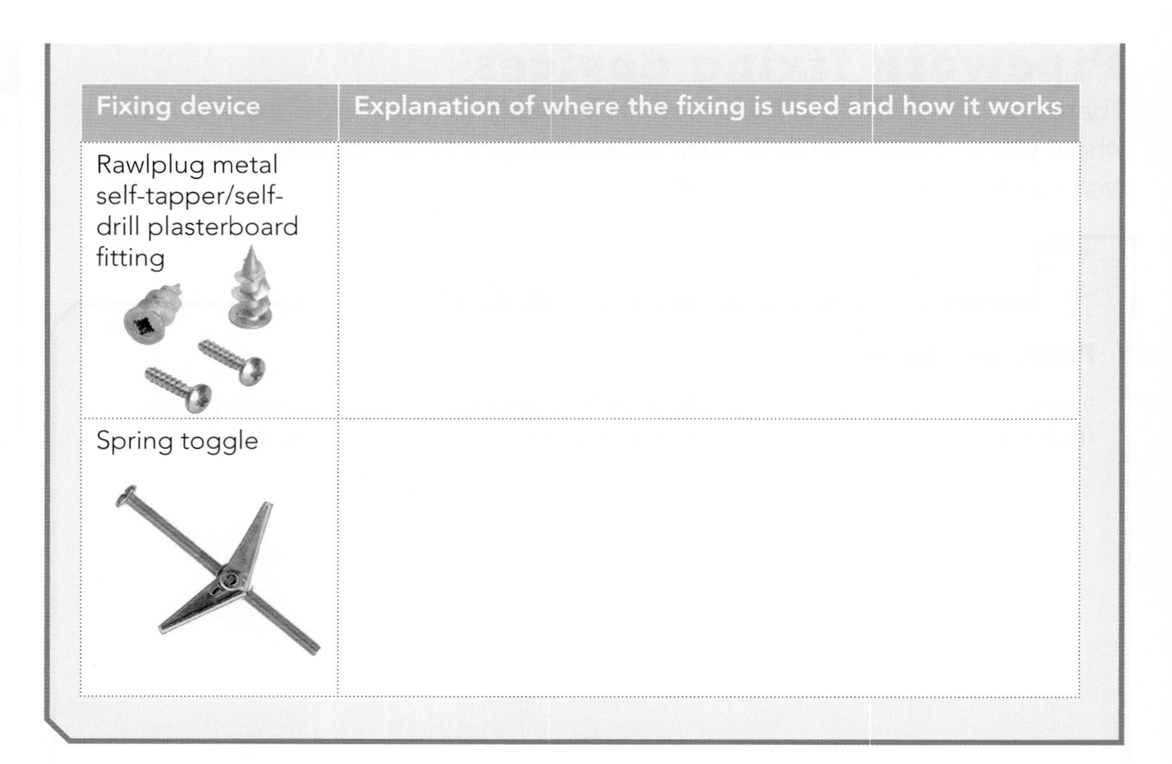

Fixing device	Explanation of where the fixing is used and how it works
Rawlplug metal self-tapper/self-drill plasterboard fitting	
Spring toggle	

key terms

Horizontal: going in a straight line across from side to side.

Vertical: going straight up and down.

Clips and brackets

Whether you are working on domestic or industrial sites you will always need to support your pipework. Don't forget the soil and waste pipes. There is a massive array of clips and brackets to choose from, made by many companies. You may have heard of snap clips and school board clips, also known as 'scor bors'. You may also be aware of nail clips to go under floorboards.

ACTIVITY

Fill in the missing data:

Tube sizes		Plastic pipe		Copper		Low-carbon steel	
mm	in	**Horizontal** spacing between clips/ brackets in metres (H)	**Vertical** spacing between clips/ brackets in metres (V)	H	V	H	V
15	½	0.6		1.2		1.8	2.4
22		0.7	1.4		2.4		3.0
28	1	0.8	1.5	1.8			3.0
35	1¼		1.7	2.4	3.0	2.7	3.0
	1½	0.9	1.8		3.0	3.0	
54		1.0		2.7	3.0	3.0	3.6

Preparing building fabric for installation

Before the installation of any appliance or pipework, you will need to follow a certain process. Above we have considered the spacing for clips and brackets to ensure something is fixed and steady. We also looked at the different plugs and screws to use in different circumstances. Here, we will look at situations where the pipework will not be visible once it has been installed. To accomplish this you will need to prepare the fabric of the building.

ACTIVITY

In groups, discuss situations where you will not be able to see your work once it has been completed.

ACTIVITY

Use the Building Regulations to assist you in this task. You may find the information on the internet or your centre may have a copy of the relevant parts of the regulations you need.

Follow the boxes from left to right, filling them in with the answers you think are relevant. In the 'Things to consider' column you will need to think about the following: Will the pipes need to be **sleeved** going through a wall? Do Building Regulations apply? How can we potentially protect pipework in a **chase**?

Job required	Tools required	Safety check on tools	PPE required	Things to consider
Cutting a trap in floorboards				
Notching joists				
Chasing walls				
Making holes through walls				

key terms

Sleeved: term to describe keeping the main system pipework away from a material which may cause damage by means of using a larger diameter 'sleeve' around it.

Chase: channel created to run pipework in a wall or a floor to hide that pipework from the owner's view.

What is meant by the following terms?

1. Making good
2. Fire stopping
3. Pre-fabrication of pipework
4. Installing pipework *in situ*
5. Protect the building fabric from damage
6. PAT testing

Identifying common symbols

A very important document you will need to be able to understand when installing pipework is the service plan. This is a plan drawn up by the architect and shows where the pipework and appliances are going to be installed. Such plans use standard symbols and colour coding, which you must be able to recognise and understand.

ACTIVITY

Symbols

Using information from the internet, colleagues and books, complete a diagram for each symbol in the space below.

Direction of flow	Isolating valve	Termination point	Drain valve	Capped end

Pressure reducing valve	Strainer	Wheel head valve	Lockshield	Single check valve
Double check valve	Circulating pump	Gas meter	Water meter	Radiator
3 port valve	2 port valve	WC close coupled	Kitchen sink	Basin

In domestic plans, colour coding is very simple: Cold = Blue, Hot = Red and Gas = Yellow. British Standards in the industrial sector are more complex.

Coding pipework

Use the internet to find out and then discuss how pipework is coded in an industrial situation. Things to consider: colours, width of bands, pipework under floors, etc. Do the rules stay the same as for the domestic sector?

key terms

Commissioning: the process of finishing a job, ensuring everything works and getting the system to operate at its optimum for the client.

Testing systems before commissioning

Once you have installed your system, you need to complete two important steps before you **commission** the system fully:

- visual inspection of fittings and connections
- pressure test the system to performance standards.

ACTIVITY

When visually inspecting a system you will not only check the fittings but also the clips and brackets, etc. In the different systems below what should you check, specific to the materials used?

1. Central heating system – copper

2. Bathroom installation waste pipework

3. LCS heating system

4. Pipework installed under floors using pushfit fittings

5. Guttering system

Hydraulic testing

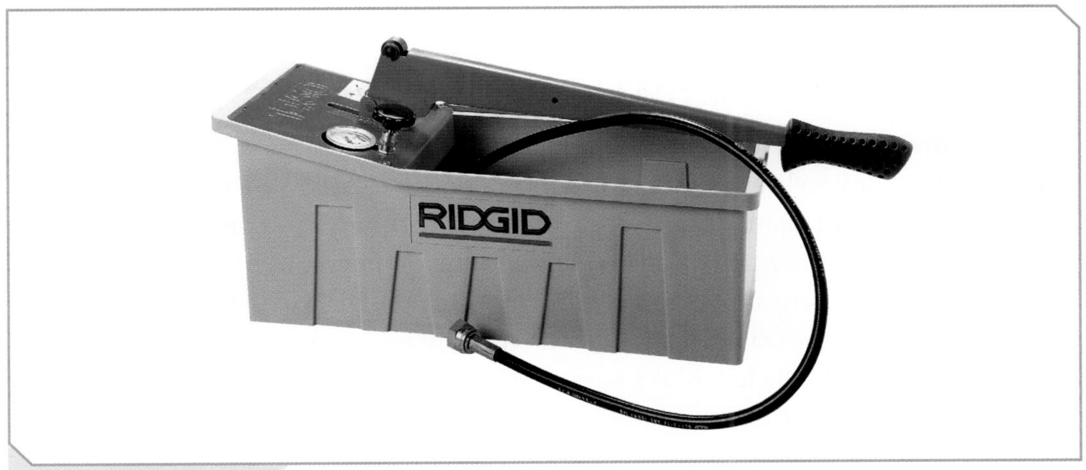

A pressure tester

ACTIVITY

Hydraulic tester

In groups, discuss what you think is the principle of a hydraulic testing machine.

ACTIVITY

State the correct testing procedure for the following materials:

Material	Brief description of how to test and correct pressures
Metallic pipework	
Plastic pipework	

Your questions answered...

I still don't understand the Building Regulation for notching joists. Sometimes when we have to drill through the joist for plastic pipework, I'm not convinced I have done it correctly.

To deal with notching joists first; the Building Regulations are to be adhered to at all times. They will set out the correct procedure to follow so the structure of the house is not damaged. There may be times when you come across a situation where a previous tradesperson has cut so much out of a joist that it could be deemed unsafe. Wood can only take so much strain. If you remove a large section of wood from a joist, then it could be weakened to the point where it will eventually fail. Make sure that you don't do this. See the diagram below for the rules applying to notching joists.

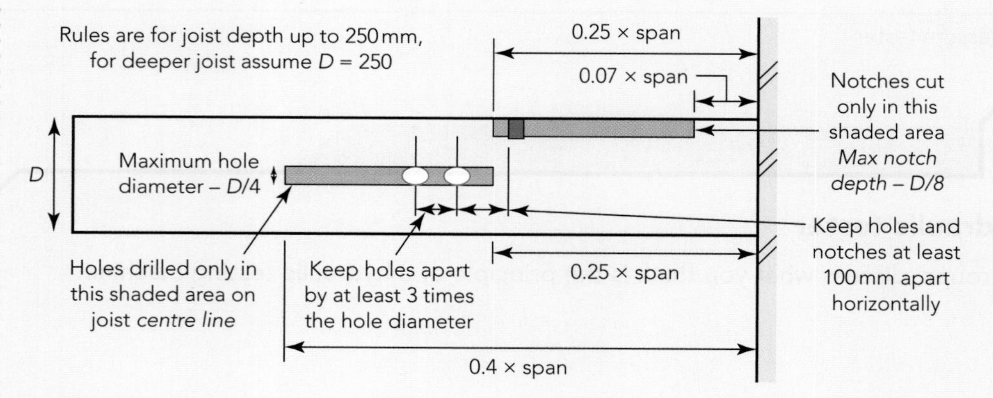

Notching joists

Let's assume we have a joist that is 3 m long (span) with a height (*H*) of 200 mm.

To work out the maximum depth (*D*) of a notch we must consider the rule of:

$$\frac{H}{8} = D$$

$$\frac{200}{8} = 25\,\text{mm}$$

Therefore our notch can only have a maximum depth of 25 mm.

We then need to consider where on the joist we can put the notch. There are rules for this too.

To work out the minimum distance from the wall we must follow this equation/rule:

Our span is 3,000 mm long so our calculation will be 0.07 × 3,000 mm = 210 mm away from the wall.

(0.07 is a constant.)

We also have to work out the maximum distance from the wall where we can notch, using the following calculation:

0.25 × 3,000 mm = 750 mm

The rules for drilling holes are :

- keep holes and notches at least 100 mm apart
- keep the holes apart by at least three times the hole cut
- holes to be made in the centre of the joist.

Where we had minimum and maximum distances with notching, we have the same situation with holes.

They are:

Minimum – $0.25 \times 3{,}000 = 750$ mm

Maximum – $0.4 \times 3{,}000$ mm $= 1{,}200$ mm

The maximum size of hole in mm is calculated as $\frac{H}{4}$

So our new calculation, given the height of the joist, gives us:

$$\frac{200}{4} = 50\,\text{mm}$$

QUICK QUIZ

1. What is the most common grade of copper used by the plumbing industry?
2. What does LCS stand for?
3. Where would solvent cement be used?
4. If you want to hide a screw, what type would be best to use?
5. What fixings might you consider for use on poor surfaces?
6. What is the minimum distance for a 2-mm horizontal capper?
7. What type of bracket should be used for LCS pipework?
8. What PPE should be worn when making a fixing for clips?
9. What are the correct tools to notch a joist?
10. How many times the working pressure should you test at for plastic pipework?

Unit 006/206

Understand and apply domestic cold water system installation and maintenance techniques

We take for granted the simple process of turning on a tap and having cold, wholesome water present. In this unit we will be looking at cold water in more detail and the regulations that must be followed in order to ensure safe and reliable supply.

The Water Supply (Water Fittings) Regulations 1999 are there to prevent contamination of a water supply, the waste of water, the misuse of water and to prevent undue consumption and erroneous measurement. The approved contractor's scheme, in association with WRAS, certifies that work is in compliance with the regulations. You may find it useful to become an approved contractor. Remember, whether or not you are qualified or certified, if you trade as a plumber you can be prosecuted for failing to meet these regulations.

You must also follow BS 6700. Ask your centre to see their copy of the document for reference.

Learning outcomes

- Know the cold water supply route to dwellings
- Know the types of cold water system and their layout requirements
- Know the site preparation techniques for cold water systems and components
- Be able to apply site preparation techniques for cold water systems and components
- Know the installation requirements of cold water systems and components
- Be able to install cold water systems and components
- Know the service and maintenance requirements of cold water systems and components
- Be able to service and maintain cold water systems and components
- Know the decommissioning requirements of cold water systems and components
- Be able to decommission cold water systems and components
- Know the inspection and soundness testing requirements of cold water systems and components
- Be able to inspect and soundness test cold water systems and components

Key knowledge

- Supply routes, systems and layouts
- CWSC and protection against contamination and frost
- Valves and controls
- Understand installation of systems and procedures to follow for safe installation
- Decommission systems

ACTIVITY

Do you know what cold water system you have in your own house or flat? When you return home from college investigate which system you have. Make notes and think of some questions to ask your tutor and employer.

Cold water supply routes, systems and layouts to dwellings

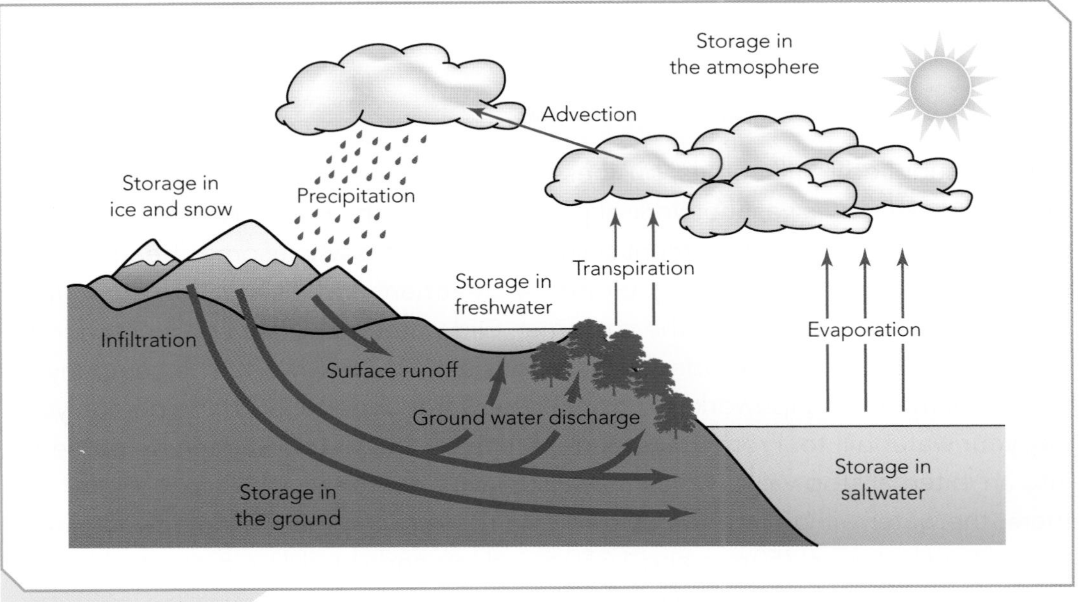

The water cycle

Look at the diagram on the previous page, which illustrates the water cycle. Here you can see how water is constantly on the move. Water in the atmosphere condenses and falls as rain or snow. The water flows into streams and rivers and ultimately the sea, or through the layers of rock beneath the surface, before evaporating back into the atmosphere. We have already discussed the process of the water passing through different rock layers, dissolving minerals that make the water hard or soft as it travels along (see page 68). But passing through these layers will also clean the water from the particles it has picked up from the air during the process of falling as rain. This is why spring water and well water is often very clean and safe to drink. Water from other sources, however, will need to be cleaned before it can be drunk or enter into the mains water supply.

THINK

ACTIVITY

Complete the table below.

Water source	Classification: **wholesome/** suspicious/ dangerous	Contamination level: **potable** and palatable/ moderately palatable/non-potable	What the water can be used for: drinking/flushing WCs only/outdoor use only
Reservoirs			
River water			
Deep well			
Shallow well			
Spring			

key terms

Wholesome (water): water of good quality, fit for humans to drink.

Potable (water): water that is safe to drink. Water doesn't have to be palatable to be potable.

Palatable (water): water that can be drunk and tastes pleasant.

Meter (water): device that records the amount of water used.

Water is supplied to all homes, businesses, etc. from sources, such as reservoirs or groundwater, by a vast network of pipes known as the mains. However, first the water is treated in water treatment plants to ensure it is wholesome. Treatment will be a combination of filtering, biological treatment and disinfection as required before the water is passed to the customer. Old water mains are constructed from cast iron, while the more modern installations use PVC pipework. The company that owns this main pipework is known as the water undertaker or the person you pay your water bill to. From the undertaker's main, the water will then be passed into an external stop valve, which may also have a **meter** attached to it. From there, the water will go into the dwelling via the service pipe. This pipework needs to be fitted between 750 mm and 1,350 mm deep and should be protected from potential damage, for instance by sheathing.

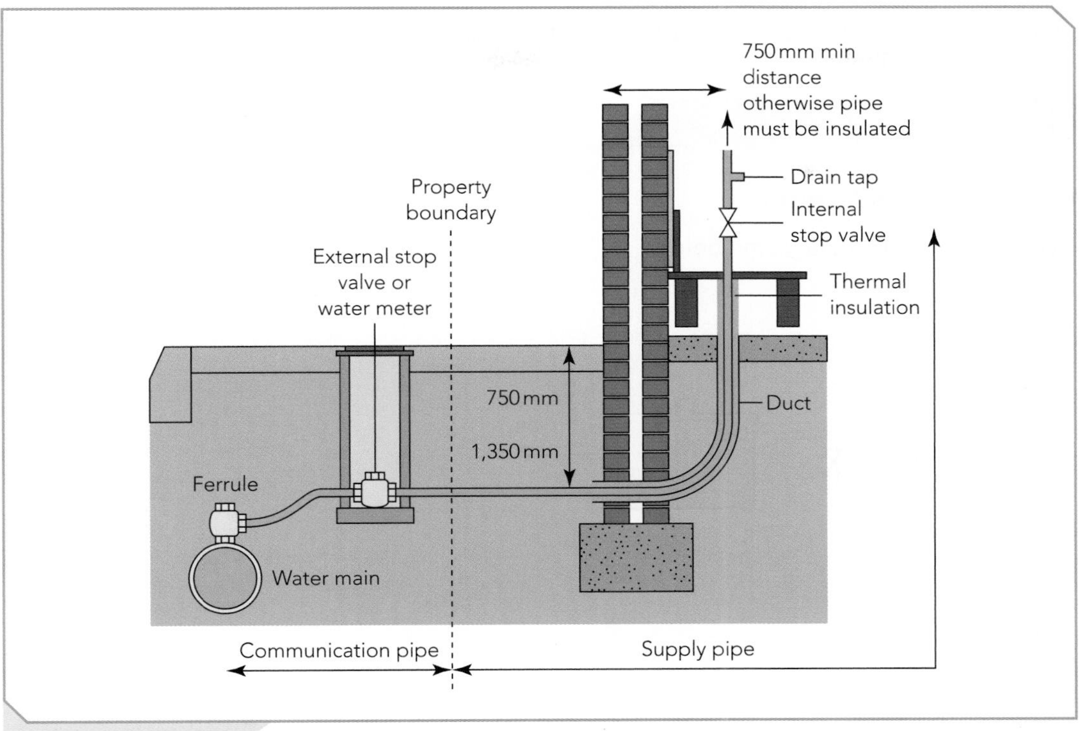

Mains service

ACTIVITY

Why do you think it is important to adhere to the measurements for the depth of pipework quoted on the previous page?

Direct cold water system

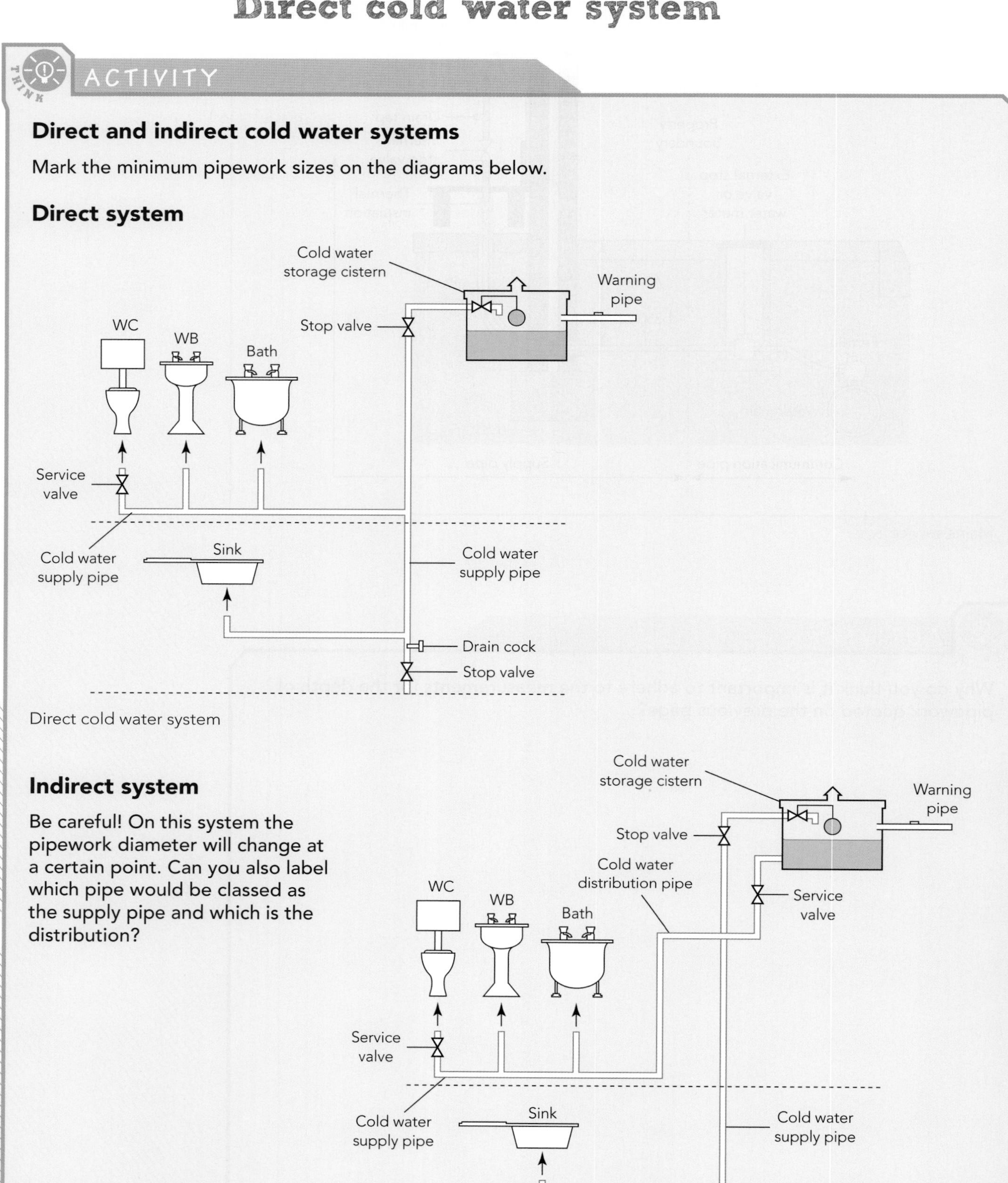

ACTIVITY

Direct and indirect cold water systems

Mark the minimum pipework sizes on the diagrams below.

Direct system

Direct cold water system

Indirect system

Be careful! On this system the pipework diameter will change at a certain point. Can you also label which pipe would be classed as the supply pipe and which is the distribution?

Indirect cold water system

ACTIVITY

Direct and indirect cold water systems

In small groups, discuss the advantages and disadvantages of both systems. Think from the point of view of the installer and the 'client'.

Direct cold water system		Indirect cold water system	
Advantages	Disadvantages	Advantages	Disadvantages

CWSC and protection against contamination and frost

Cold water storage cisterns (CWSCs) are the main storage containers to be found in domestic dwellings. Do not confuse them with tanks. Tanks belong to the army and have nothing to do with plumbing, but you may hear the term used quite a lot. WRAS and the Water Regulations state that the cistern must be fitted to a particular standard to ensure the water does not become contaminated. This requires the cistern to have certain fittings.

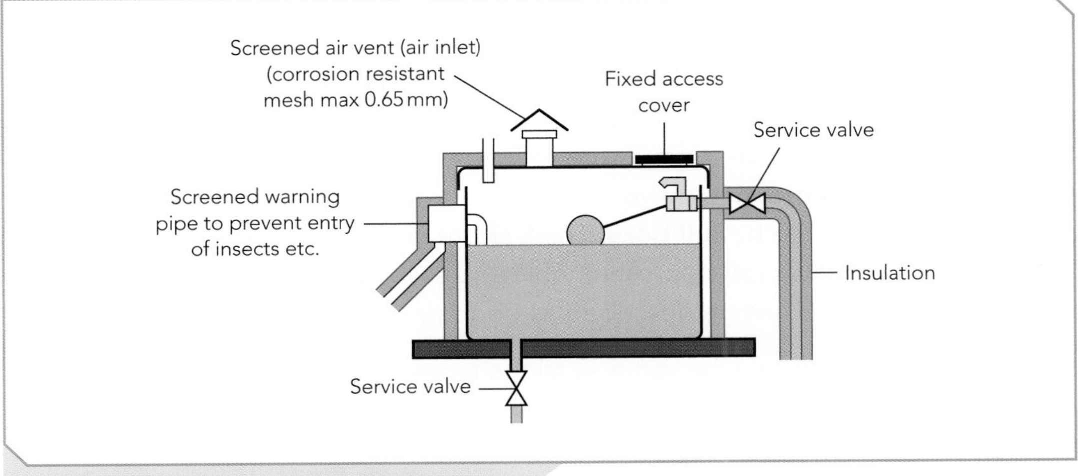

CWSC contamination prevention measures

ACTIVITY

Look at the diagram on the previous page and then answer the following questions.

1. Why is an air vent (air inlet) fitted?

2. Screens need to be fitted in the vent and the warning pipe connection – what is their purpose?

3. What stops the cistern from freezing over or heating up?

4. On to what type of base must the cistern be fitted?

5. Where the vent pipe enters the cistern at the top, what must be fitted to create a seal?

Valves and controls servicing and maintenance

There are many valves that can be used on cold water systems. What can be confusing is the terminology used to describe service valves and isolation valves – the activity below will help you with this.

Isolation valves are used to isolate a component for various reasons. The regulations do not state that isolation valves have to be fitted before every termination point. However, if you were to fit these under the bath, for instance, this would be **good practice** on your part. Be careful when fitting them on gravity systems as they have to be **full bore**. Think about when you have had to change a tap and found isolation valves present. This also saves water. That's the reason why fitting isolation valves can be described as good practice.

Service valves look exactly the same as all the types of isolation valves you use on a daily basis. It is the name that changes. When fitted before a float valve, the isolation valve is known as a service valve. This is important to remember for assessments.

key terms

Good practice: a term used to describe work that is not regulated but will assist with saving time, effort and resources in the future.

Full bore: a fitting or valve that has the same diameter as the pipework it is fitted to, so the flow is not restricted.

ACTIVITY

For this task you will need to recognise and understand the use and purpose of the various valves described and explain how to maintain and service them. Knowing the terminology of the working parts will assist in your description. We have also put termination points where the water will come out, as these will need to be serviced at some point. If the valve or termination points are known by other names, these are included also.

We have done one for you to help you complete the task.

Valve	Also known as	Installations where the valve can be used	How to maintain/service
Float valve	Ball valve		
Outside tap	Bib tap		
Non-return valve	Check valve	*Fridge/outside tap/ showers/mixer taps/bidets*	*Take valve out and check operation spring. Clean out any grit present. Refit and test the back flow operation by charging and pressurising the pipework in front of the directional arrow. Release pressure at the back of the valve to see if water returns*
Isolation valve	Ball 'o' fix		
Wheel head valve	Gate valve		
Drain off	DOC		
Tap			

ACTIVITY

ASK

In groups, find out which FOUR different float valves are available and the BS numbers associated with them. Don't forget the Torbeck valve.

Safe installation procedures

In this section you will be given a project to complete. Using all the information you have obtained from the units including this one and all the practical knowledge you have obtained so far, you will be asked to consider a scenario for an installation as a plumber. This will also be strongly linked to the on site assessments that you will need to complete.

ACTIVITY

Complete the table on pages 126–7. PlumbBob Squared Pipes has just taken on a job installing a new cold water system.

You will use the drawing below showing the type of system you have been asked to install. You are installing the system from new at the stop valve/cock. You will be required to run some pipework across joists for the bathroom. All pipework must be clipped correctly and the customer, Michael Mangan, has stated that no plastic pipework must be used. You must tell the carpenter, Amy Wilson, what base to put the CWSC on.

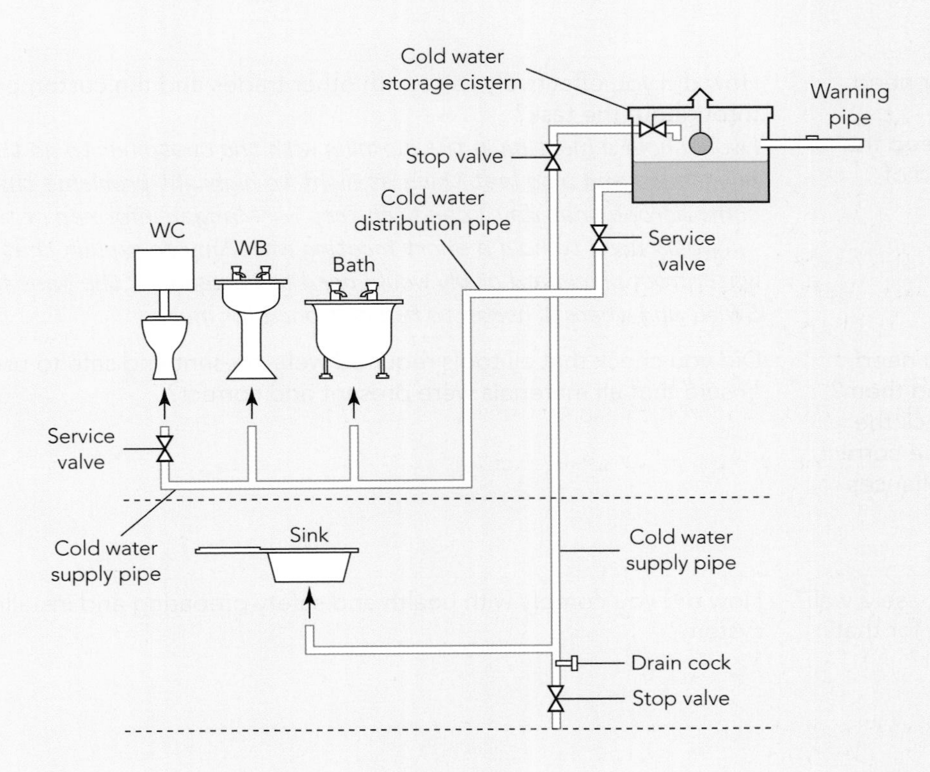

Michael has stated he is on a tight timescale and you have agreed to complete the installation in two days, as the sanitaryware is already installed and the customer has asked you to install a stored cold water supply to the kitchen sink.

In the third bedroom where the pipework rises up, the wall is made of plasterboard.

Complete the table on pages 126–7 highlighting the key issues and how to deal with them. The first column gives you the key issues that need addressing. In the second column, explain how you would complete the works.

One question has been answered for you as an example.

Key issues you need to mention in your write-up	Action taken
What are the documents you would use to aid you with the installation?	How did you check that all necessary plans, specifications and verbal instructions were available before starting the job?
Does the plasterer need water for his mix? Do you need to keep the customer informed of progress?	How did you effectively liaise with other trades and the customer throughout the task? *I would hold a meeting in the morning with the customer to go through any issues and also last thing at night to highlight problems that we have come across that day. I can then keep Mr Mangan informed of timescales.* *I would need to hold a short meeting with Amy to explain that waterproof chipboard or ply would need to be used for the base for the CWSC and where it needs to be positioned for me.*
What tools do you need, have you inspected them? How? Did you check the materials list for the correct materials and appliances needed?	Did you check that all tools required were present and safe to use and ensure that all materials were present and correct?
Did you have to chase a wall? Did you need PPE for that? Are walkways clear?	How did you comply with health and safety preparing and installing the system?
Did you remove all the ornaments in the way? Dust sheets laid?	How did you take precautions to protect the customer's property from damage?
Have you notched joists? If so, you will need to follow certain regulations.	How did you prepare the building fabric to receive pipework?

Key issues you need to mention in your write-up	Action taken
Correct materials, jointing process? Clip distances?	How did you install the main system pipework to industry standards?
Have you connected all the correct WRAS fittings?	How did you install the CWSC to industry standards?
Did you go to another call-out? Did you cap the pipework?	If needed, how did you take precautions to prevent the use of unfinished installations?
Have you soldered all the fittings, tightened the DOC?	How did you carry out checks on fittings and fixtures? Please explain.
How did you test the system?	How did you soundness test the system in accordance with industry standards?
Do you need to protect the pipework in the loft?	Did you fix insulation as required; if so what type and where?
Have you checked the operating procedures of all the valves?	How did you commission the system and pass over the completed job to the supervisor and customer? What information did you give?

Decommissioning cold water systems

Describe what steps you would take to decommission the following systems.

1. Full drain-down of a direct system. The customer has other trades in the dwelling who need water and don't know that you will be draining down.

2. Decommission the distribution pipework only.

Your questions answered...

What is stagnant water? Is it dangerous?

Stagnant water is standing water where there is no throughflow. Stagnant water can be dangerous for drinking because it provides a better incubator than running water for many kinds of bacteria and parasites. Standing water can be created in two main ways in a domestic dwelling.

The first is related to the CWSC and where the distribution pipe is fitted in relation to the float valve. You should always fit the distribution pipe to the opposite side of the float valve; this will keep a constant motion of water going through the cistern, preventing it becoming stagnant. Otherwise the water will always find the quickest route, which may lead to the furthest area from the float valve becoming stagnant as there is no through flow of water in that area.

The second can be created by the existence of a dead leg. This is when a piece of pipe of some length is not needed any more and is capped to stop it being used, leaving water in it, which has no outlet. You must always remove dead legs and instead cap the dead pipe as near as possible to the running pipework in constant use.

QUICK QUIZ

1. What process is involved in water in the sea entering the atmosphere to become a cloud?

2. Is spring water generally safe to drink? What about rainwater?

3. What is the minimum depth for a service pipe entering a dwelling?

4. What is the minimum sized pipework for cold water mains?

5. What is the minimum size for a distribution pipe on an indirect system going to a bath?

6. Which 'part' float valve should be used on a CWSC?

7. What is the BS number for installation and design of cold water systems?

8. What is one disadvantage of just having cold water?

9. What is a termination point?

10. Why would you use a cap end?

Unit 007/207

Understand and apply domestic hot water system installation and maintenance techniques

One of life's little luxuries is being able to turn on a tap, wait a few seconds and have warm water flowing for as long as you need it. In this unit you will be considering how to produce this hot water. You will be working with many systems on a day-to-day basis that use different methods to heat the water and different energy sources, e.g. gas, oil, solid fuels, electricity and solar.

The terminology for the two main system types must not be confused with that used in cold water systems:

- direct systems – when the water is heated directly from the heat source
- indirect – when the heat source is separated from the hot water and transfers the heat by another means.

Again, BS 6700, Water Regulations and Part L must be followed. Ask your centre to see their copy of the documents for reference.

Learning outcomes

- Know the types of hot water system and their layout requirements
- Know the site preparation techniques for hot water systems and components
- Be able to apply site preparation techniques for hot water systems and components
- Know the installation requirements of hot water systems and components
- Be able to install hot water systems and components
- Know the service and maintenance requirements of hot water systems and components
- Be able to service and maintain hot water systems and components
- Know the decommissioning requirements of hot water systems and components
- Be able to decommission hot water systems and components
- Know the inspection and soundness testing requirements of hot water systems and components
- Be able to inspect and soundness test hot water systems and components

Key knowledge

- Convection and stratification
- Hot water systems and layouts
- Showers, valves and controls
- Installation of pipework for showers from CWSC
- Understand installation of systems and procedures to follow for safe installation
- Decommission systems

ACTIVITY

Do you know what hot water system you have in your own house or flat? When you return home from your centre investigate which system you have. Make notes and think of some questions to ask your tutor and employer.

Convection and stratification

We have already talked about **convection** in Unit 204, and if you can remember it is linked closely to Boyle's law, which states that when air is heated the molecules will move further apart and rise. In hot water this process is also called convection where the water that gets heated rises as it becomes less dense. When the water cools it will 'fall'. Another process that happens in this context is **stratification**.

key terms

Convection: the movement of molecules within fluids (i.e. liquids, gases). It is also known as a mode of heat transfer.

Stratification: the different layers of temperatures that form in the hot water cylinder from cold at the bottom to hot at the top.

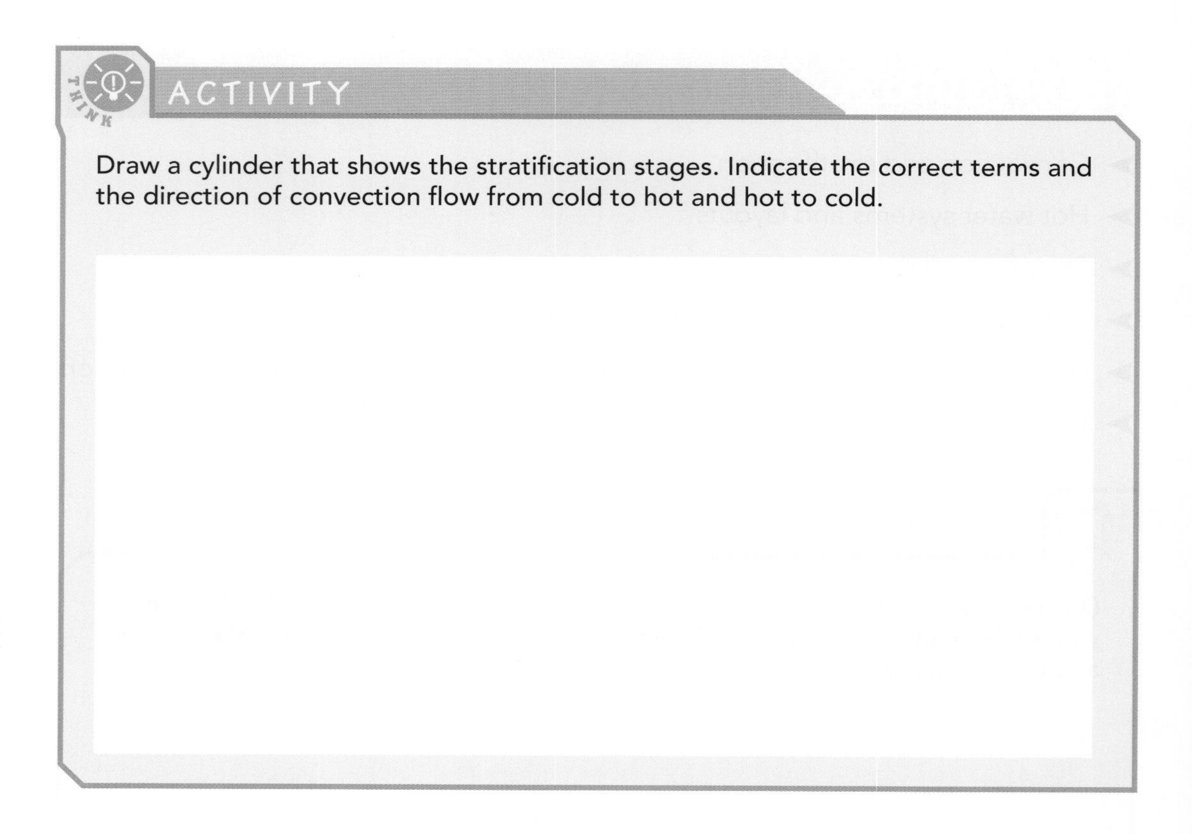

ACTIVITY

Draw a cylinder that shows the stratification stages. Indicate the correct terms and the direction of convection flow from cold to hot and hot to cold.

Hot water systems and layouts to dwellings

The diagram below shows the numerous hot water systems that you need to understand. There are two categories to choose from: centralised and localised.

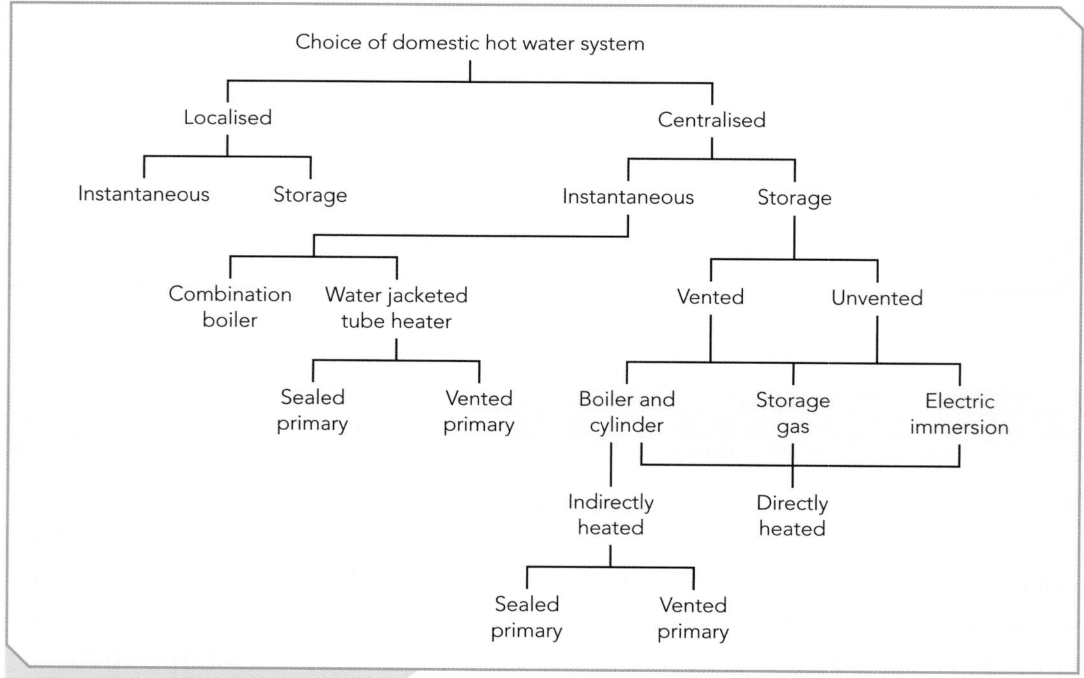

Domestic hot water system

- Centralised systems – where the water is heated and can be stored centrally if required.
- Localised systems – where the water only needs to be produced in a certain place and therefore can be heated directly at that point.

ACTIVITY

Complete the table below.

Which of the systems is classed as centralised or localised? Tick the correct box.

System	Centralised	Localised
Condensing combi boiler		
Indirect hot water		
Unvented hot water		
Single point hot water		
Fortec		

key terms

Fortec: a brand name of a combination cylinder where the CWSC sits directly on top of the cylinder to save space. You will hear the term Fortec more than the term combination cylinder.

ACTIVITY

In groups, discuss and list different systems you have seen since you started your apprenticeship.

Direct hot water systems

Direct hot water system

On the diagram below, mark the minimum-sized pipework for all pipework present. Label where the cold feed is.

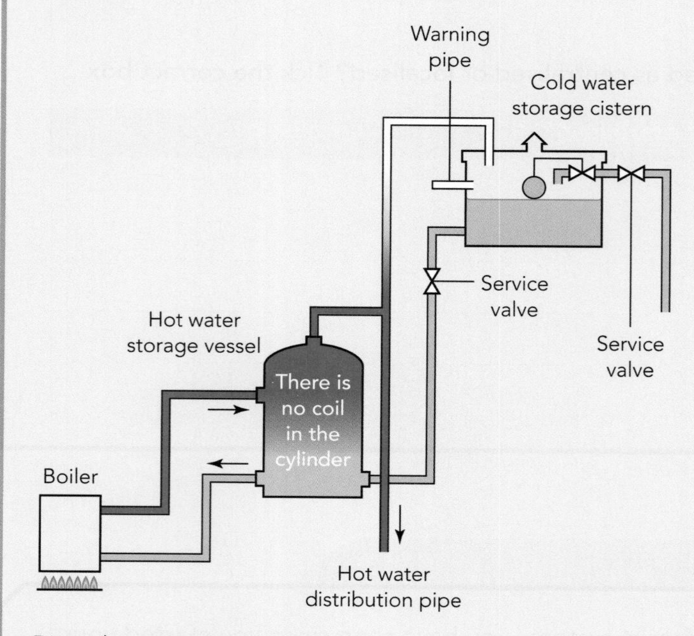

Direct hot water system

ACTIVITY

The direct hot water system above works by convection of the water; the term used in the industry is a **gravity system**. Answer the questions below regarding direct systems.

1. Can you name another way of heating water directly in the cylinder?

2. What is an Economy 7 system?

3. For the system in the diagram to work correctly using convection, what size pipework should you use to connect the boiler to the cylinder?

4. What is the hot water pipework coming from the top of the cylinder to the CWSC known as?

5. What is the hot water pipework to the termination points known as?

6. What is the minimum sized pipework to the bath?

Indirect hot water

As you can see in the diagram below, the system has been joined by a feed and expansion cistern and the water is separated from the boiler by means of a **coil**. Therefore, the water in the cylinder is now being heated indirectly.

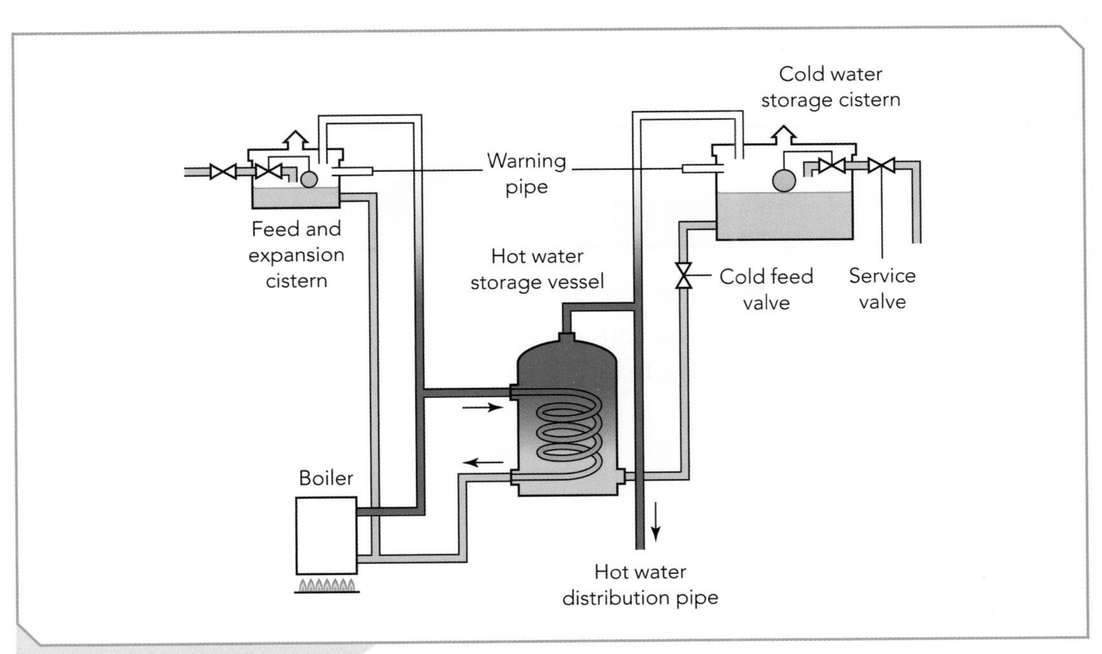

Indirect hot water system

key terms

Gravity system: a system that works without the requirement of pumps, using only gravity and atmospheric forces.

Coil: heated copper heating element in a cylinder fitted in circular fashion so the heat transfer can be distributed evenly throughout the cylinder.

ACTIVITY

Describe the working principles of the indirect hot water system in the space below.

ACTIVITY

Direct and indirect hot water systems

In small groups, discuss how you would control the temperature of direct and indirect hot water systems. Write down your conclusions.

It is important that the temperature does not exceed 65 °C. In your groups, discuss and give TWO reasons why this is.

1.

2.

ACTIVITY

Refer to Unit 204 if you need help with this activity.

Your customer requires a cylinder with a capacity of 150 litres. What are the dimensions of the cylinder going to be?

Show your workings below.

Other hot water systems

Complete the table below using the internet and other literature to assist.

We have completed one row for you.

Hot water system	Brief description of its basic working principles
Combination storage system	
Instantaneous multi point	
Combi boiler	
Water jacketed system	*Uses the cylinder to act like a radiator. It heats up the contents of the cylinder with the cold water pipework passing through it heating up as it passes. Similar to how the indirect hot water system works, only reversed*
Single point	
Indirect single feed	

Showers, valves and controls

Some of the valves that are used in hot water systems are also used in cold water systems. In the previous unit you completed a similar task to the following on these valves.

There are also specialist valves used in hot water systems, which you will need to understand, together with various shower types.

ACTIVITY

For this task you will need to know and understand the valves described and explain how to maintain/service them.

Valve	What is the purpose of the device and where is it used?	How to maintain/service
Strainer		
Thermostatic mixer valve		
Secondary pump		

ACTIVITY

Showers

For this task showers of different types are listed below. Describe their working processes and the limitations of system pipework in the space provided.

Shower type	Describe the working process. Include if equal pressures are required
Electric	
Power shower	
Thermostatic mixer	
Mixer	

Connecting pipework for showers to CWSC

ACTIVITY

In pairs or individually, draw a complete indirect hot water system and label all the valves and pipe sizes correctly.

You need to install a mixer shower in the system. You will need to show the correct positions of the pipework connections to the CWSC. Be careful! You do not want to scald the customer.

Safe installation procedures for hot water systems

In this section you will be given a project to complete. From all the information you have obtained from the units, including this one, and all the practical knowledge you have gained so far, you will be asked to consider a scenario for an installation as a plumber. This will also be strongly linked to the on site assessments that you will need to complete.

key terms

Primary flow and returns: the pipework from the boiler to the cylinder.

ACTIVITY

MANAGE

Complete the table on the next two pages. PlumbBob Squared Pipes has just taken on a job installing a new hot water system.

You will use the drawing below showing the type of system you have been asked to install. You are installing the system from new at the stop cock/valve. You will be required to run some pipework across joists for the bathroom, and for the **primary flow and returns**. All pipework must be clipped correctly and the customer, Denise Haffner, has instructed you to use plastic pipework under the floor to the bathroom, but she wants copper to the termination points above ground. She would also like you to fit a thermostatic mixer shower valve above the bath and you will have to chase the wall for the pipework.

In the airing cupboard the masonry is a poor surface to attach fixings for clips. Choose the correct device to use.

The CWSC is on the third floor and Denise would like to know what sort of pressure she will have in the shower, which has about 7 m head of pressure.

Complete the table on pages 141–2 highlighting the key issues and how to deal with them. The first column gives you the key issues that need addressing. In the second column, explain how you would complete the works.

Your tutor will want to see how you have dealt with these issues so be sure you have addressed them.

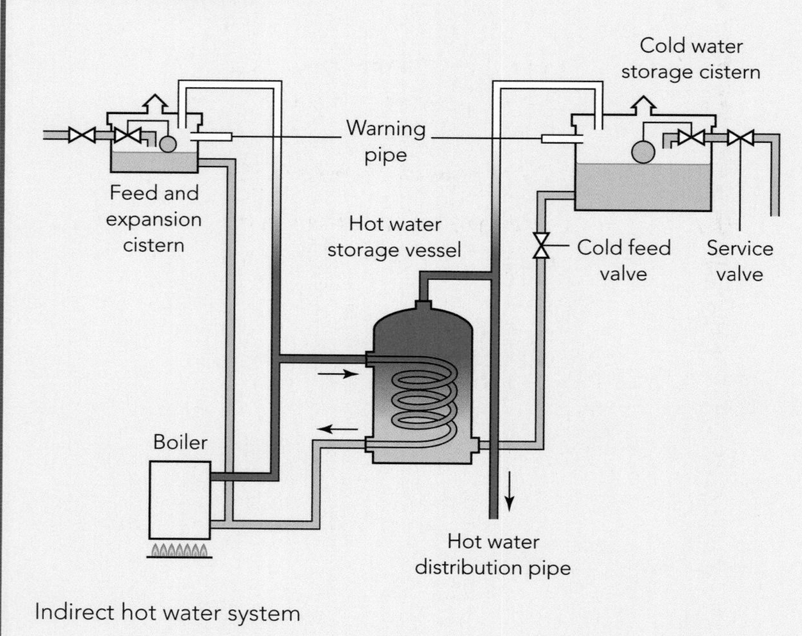

Indirect hot water system

Key issues you need to mention in your write-up	Action taken
State the documents you would use to aid you with the installation. Regulations?	How did you check that all necessary plans, specifications and verbal instructions were available before starting the job?
Do you need to keep the customer informed of progress?	How did you effectively liaise with trades and the customer throughout the task?
Did you check the materials list for the correct materials and appliances needed?	Did you check that all the tools required were present and safe to use and ensure that all materials were present and correct?
Did you have to chase a wall?	How did you comply with health and safety preparing and installing the system?
Dust sheets laid?	How did you take precautions to protect to customer's property from damage?
Have you notched joists? If so, you will need to follow certain regulations.	How did you prepare the building fabric to receive pipework?
Correct materials, jointing process?	How did you install the main system pipework to industry standards?

Key issues you need to mention in your write-up	Action taken
Have you connected all the correct WRAS fittings?	How did you install the cylinder and CWSC etc. to industry standards?
Did you cap the pipework?	If required, how did you take precautions to prevent the use of unfinished installations?
Have you soldered all the fittings, tightened the DOC?	How did you carry out checks on fittings and fixtures?
How did you test the system?	How did you soundness test the system in accordance with industry standards?
Did you carry out pressure and flow rates? How?	How did you carry out flow and pressure tests to standard?
Do you need to protect the pipework in the loft?	Did you fix insulation as required? If so, where?
Have you checked the operating procedures of all the valves?	How did you commission the system and pass over the completed job to the supervisor and customer? What information did you give?

Decommissioning hot water systems

ACTIVITY

Describe what steps you would take to decommission the following systems:

1. Full drain-down of an indirect hot water system.

2. Decommission the distribution pipework only.

3. Decommission a mixer shower but leave the house with water for the rest of the bathroom.

Your questions answered...

I have turned off the cold feed to decommission the hot water distribution pipework. When I had completed the replacement of the bath taps and tried to reinstate the hot water supply, I failed to get the hot water to come through. I know the valve works as it's a full bore lever valve. What's the problem?

This sounds like an 'air lock', which is down to poor installation. An air lock occurs where the pipework installed is going up at an angle somewhere along its route and then sharply down, usually next to an elbow; this will stop the water from pushing the air through. Always double check the valve to see if it is open. To do this you should close it first, undo the bottom nut to release the pressure and open the valve into a bucket or vessel of some type. If there is water, then you do have an air lock. Don't forget you are working on a low pressure system; this is why gravity systems must be fitted with a slight fall allowing any air to escape. There are many ways to reinstate the supply by removing the air by other means. You should see your boss for their preferred method.

QUICK QUIZ

1. How do you heat the water in a direct system, which does not have a boiler?

2. What does E7 stand for?

3. Why is 22 mm pipework used for hot water distribution pipework to baths?

4. What is stratification?

5. What is a coil?

6. What is the ideal temperature for storing hot water?

7. What is convection?

8. How does an electric shower work?

9. Can you fit mains cold water and gravity hot water to a simple mixer shower?

10. What is a Fortec system?

11. If replacing an old cylinder, what regulations must be followed to bring the system up to standard?

12. What is the BS number for hot water installations?

Unit 008/208

Understand and apply domestic central heating system installation and maintenance techniques

In this unit we will look at the vast array of heating systems and types available to you and your customers. We will also look at the valves that aid the system to maintain a constant temperature. Modern methods of installation will aid in the conservation of fuels and, although you will come across old heating systems in dwellings that do not meet the new standards, a good understanding of their principles is still required. We will not be looking at gas pipework or installation in this unit, as you need to be deemed competent under the Gas Safe registration scheme before dealing with gas installation.

BS 2000, Water Regulations and Part L must be followed for this unit. Ask your centre to see their copy of the documents for reference.

You may need to go through this unit twice.

Learning outcomes

- Know the uses of central heating systems in dwellings
- Know the types of central heating system and their layout requirements
- Know the site preparation techniques for central heating systems and components
- Be able to apply site preparation techniques for central heating systems and components
- Know the installation requirements of central heating systems and components
- Be able to install central heating systems and components
- Know the service and maintenance requirements of central heating systems and components
- Be able to service and maintain central heating systems and components
- Know the decommissioning requirements of central heating systems and components
- Be able to decommission central heating systems and components
- Know the inspection and soundness testing requirements of central heating systems and components
- Be able to inspect and soundness test central heating systems and components

Key knowledge

- Central heating systems and pipework layouts
- Heat emitters
- Heating valves, pumps and temperature controls
- Understand installation of systems and procedures to follow for safe installation
- Decommission systems

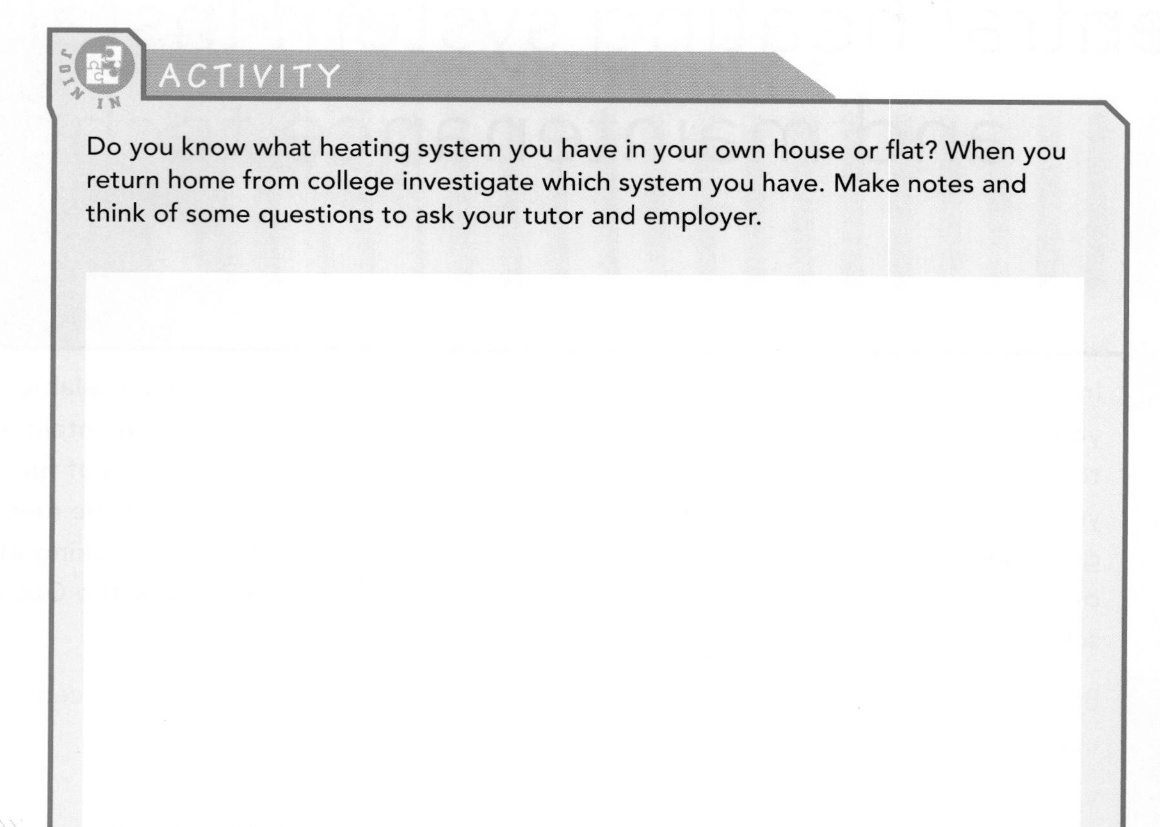

Do you know what heating system you have in your own house or flat? When you return home from college investigate which system you have. Make notes and think of some questions to ask your tutor and employer.

Central heating systems and pipework layouts

One-pipe heating systems

One-pipe systems come in two different types. One type is known as a semi-gravity system as the heating is pumped and the hot water works under gravity. Some very old one-pipe systems will also have been installed without the pump, making the system full gravity. This system is not allowed to be installed in new dwellings as it doesn't meet Building Regulations. By looking at the connections at the heat emitter to the main system pipework you can determine whether it is a one-pipe or two-pipe system.

One-pipe heating system

On the diagram below, label the minimum size for all pipework present.

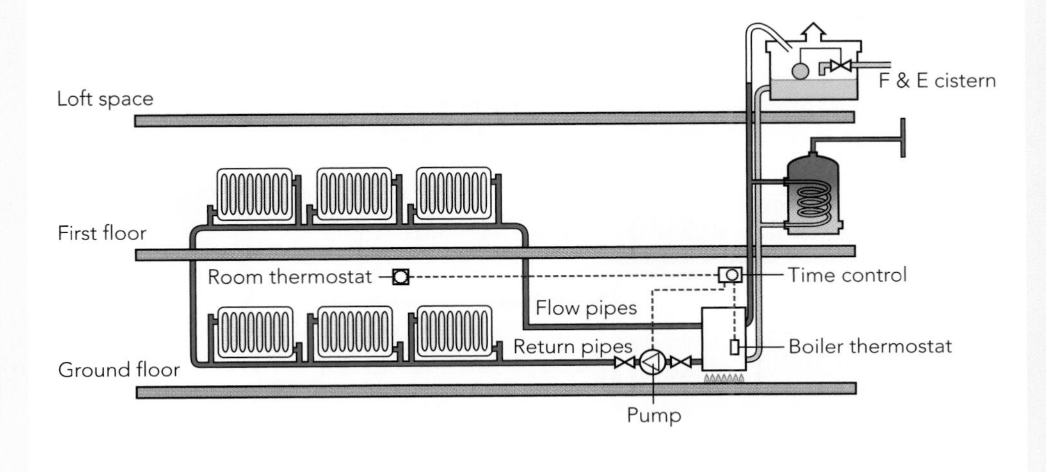

Two-pipe semi-gravity system

ACTIVITY

THINK

Two-pipe semi-gravity system

On the diagram below, label the minimum size for all pipework present.

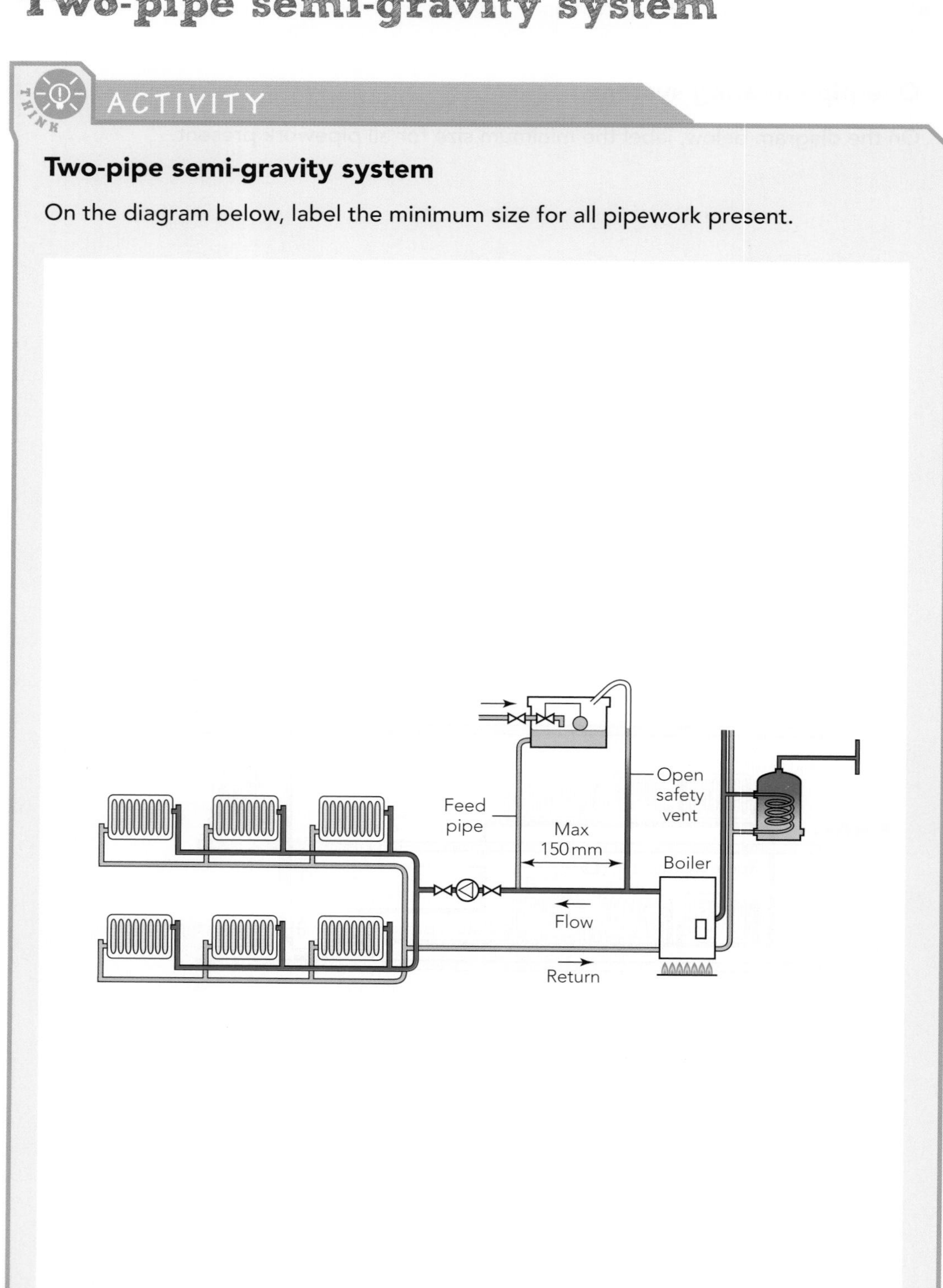

key terms

Flow and return from the boiler: these are called the primaries. The flow and return to the cylinder will now become the secondaries.

The only difference between the two systems is that the two-pipe system incorporates a **flow and return from the boiler**. It, too, cannot be installed in new dwellings.

Fully pumped system incorporating mid-position valve

THINK ACTIVITY

Fully pumped mid-position

On the diagram below, label the minimum size for all pipework present.

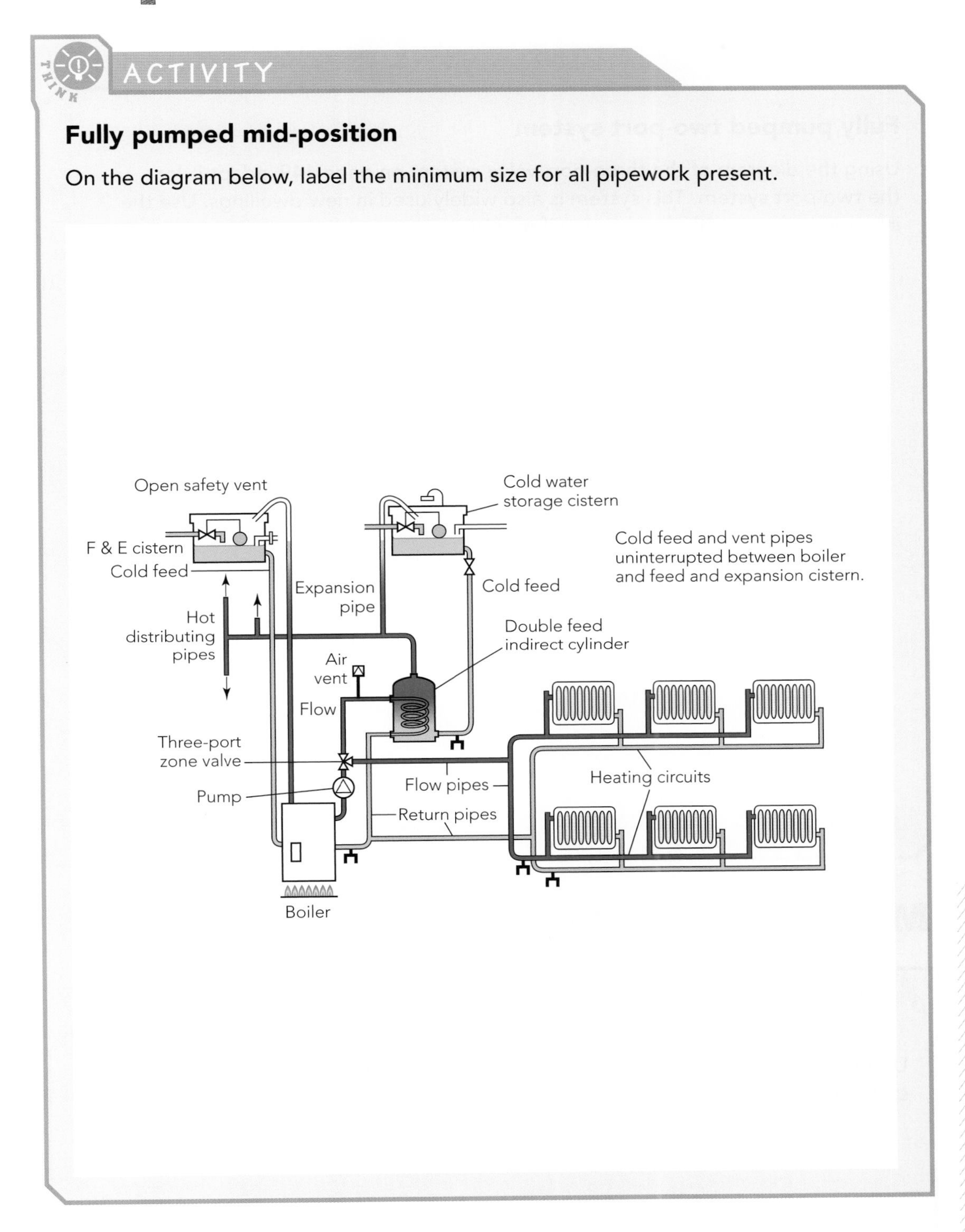

This is the system that is most commonly installed in new properties. Incorporating a mid-position three-port valve has an advantage over the old two-position three-port valve in that you can have programmed heating and hot water working simultaneously. This system fully meets the requirements of the regulations if it is also fitted with all the other fuel-saving controls.

Fully pumped system incorporating two-port valves

ACTIVITY

Fully pumped two-port system

Using the diagram of the three-port valve system on page 149, adapt it for the two-port system. This system is also widely used in new dwellings. Use the internet to assist with the drawing if required.

Micro-bore systems

ACTIVITY

Undertake some research to find out about the micro-bore system and write a short statement explaining what it is. Don't forget to mention the pipe sizes available.

Vent, pump and cold feed positioning

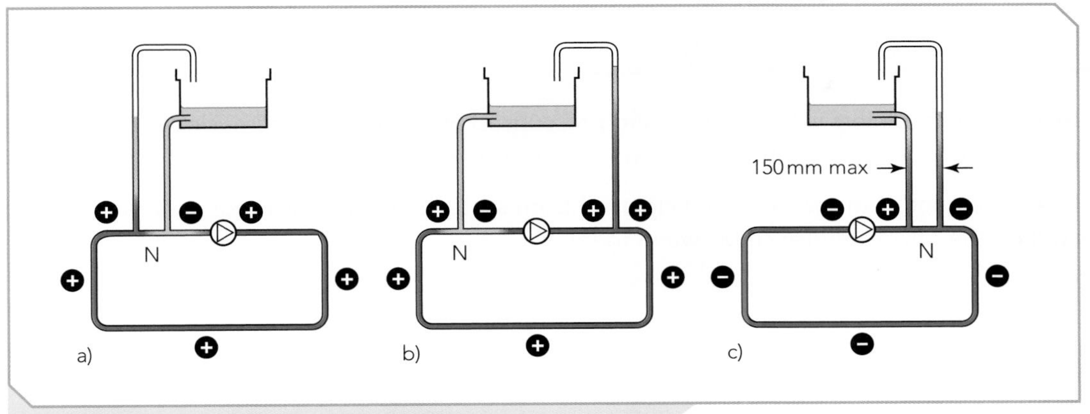

(a) Vent, (b) pump and (c) cold feed positioning

ACTIVITY

Complete the table below explaining the outcome to the system if you position the pump, vent and cold feed as shown in the diagrams above.

	Describe the possible outcome to the water in the system if the connections made are like this
A	
B	
C	

Combination and sealed system boilers

ACTIVITY

Both combination and sealed system boilers will work by using pressurised water in the system. Under Part L of the Building Regulations, you will be required to install new condensing boilers with a SEDBUK rating of A.

In the space below, write a statement of the working principles of both systems. Explain how hot water is produced by both. This will be the major difference between them.

ACTIVITY

Your customer requires a new boiler. Their old system was a three-port system. They have a traditional system boiler and do not want to change to a condensing combi.

In groups, discuss which boiler you should fit and which controls will bring the system up to standard, assuming the customer does not have them. Write up your conclusions. Do you agree with the rest of your group? If not, give your reasons.

Boiler systems

Give a brief description of the working principles of the boilers listed and explain how the **products of combustion** in each type are emitted. You should use the internet and other literature to assist you.

We have completed one row for you.

System	Description
Solid fuel boiler	
Electrical boiler	
Condensing	
Back boiler	
Oil boilers	*Oil boilers work in the same way as system and combi boilers and can incorporate all valves and controls as a normal gas system. Oil is injected and ignited to heat the system and reach the temperature required. The boiler can be wall mounted. Used in situations where you cannot get gas*

key terms

Products of combustion: unburnt and created gases that have been produced from burning fuels.

Heat emitters

The term *heat emitter* is used to describe a device that heats any area you desire.

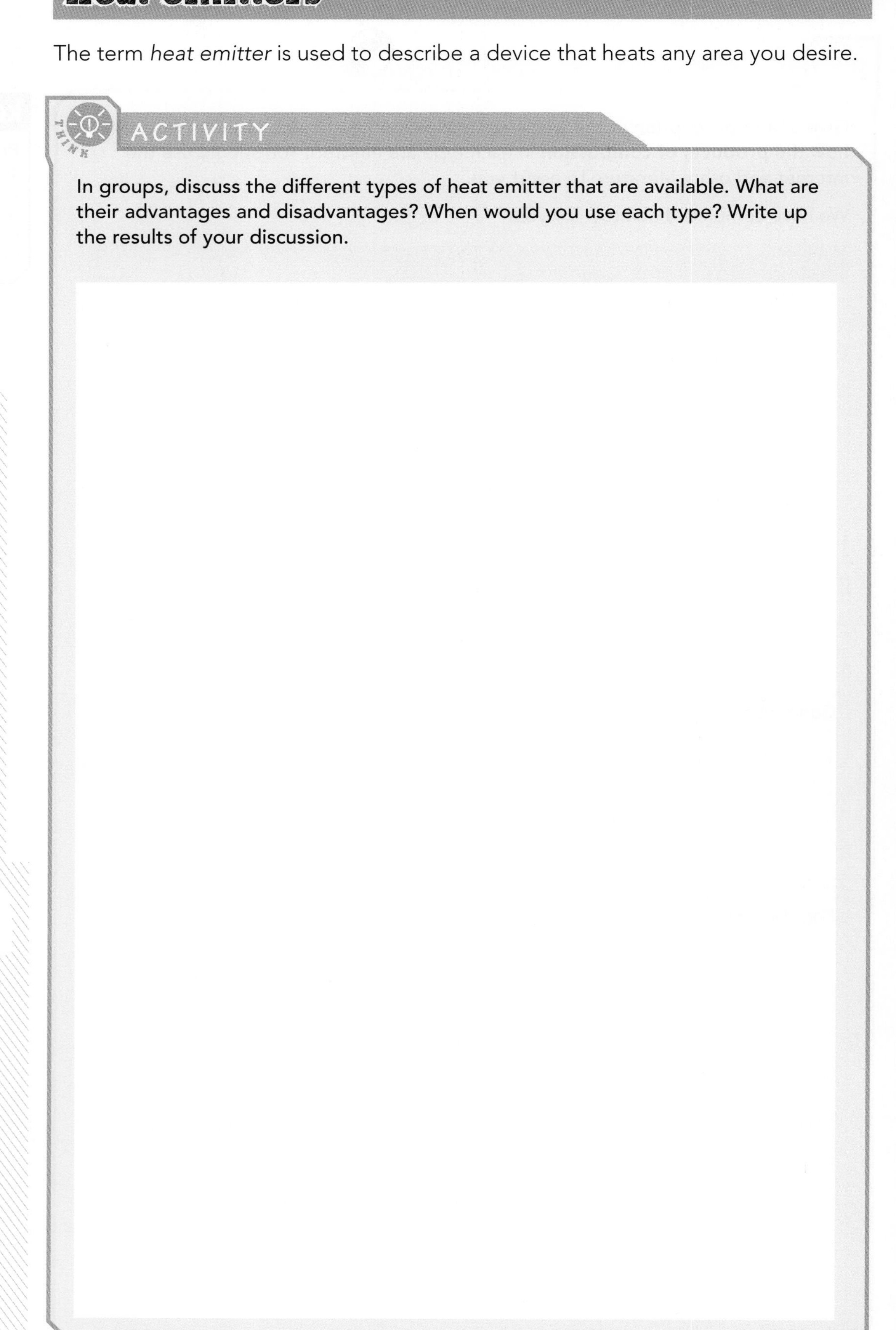

THINK
ACTIVITY
In groups, discuss the different types of heat emitter that are available. What are their advantages and disadvantages? When would you use each type? Write up the results of your discussion.

Heating valves, pumps and temperature controls

key terms

Optimum: the best possible conditions for a system to operate at its full potential.

When working with heating systems you will come across a lot of valves and controls. Each one is designed to complete a specific job, which will keep the conditions at the **optimum** for the most efficient working of the system.

ACTIVITY

For this task you will need to know and understand the different valves and controls and explain their function.

Name each of the valves or controls shown in the table below and describe their function.

Valve/control	What is the function of the valve/control?

ACTIVITY *continued*

Valve/control	What is the function of the valve/control?

Understand installation of systems and procedures to follow for safe installation

In this section you will be given a project to complete. Using all the information you have obtained from the units, including this one, and all the practical knowledge you have gained so far, you will be asked to considered a scenario for an installation as a plumber. This will also be strongly linked to the on site assessments that you will need to complete.

ACTIVITY

Complete the table on pages 158–9. PlumbBob Squared Pipes has just taken on a job installing a heating system in a large dwelling.

You will use the drawing below showing the type of system you have been asked to install. You are installing the system from new at the stop valve/cock. You will be required to run the pipework across joists to the appliances and heat emitters. All pipework must be clipped correctly and the customers, Mr and Mrs Davies, have instructed you to use LCS pipework in the boiler house. They require copper throughout the dwelling as normal.

A sealed system boiler and fully programmable controls with TRVs have also been requested.

The bathroom is already fitted and will not need modernisation.

All walls are in very good order; you will be instructed to cut the new hole through the wall for the flue and all the correct PPE must be worn.

Complete the table on pages 158–9 highlighting the key issues and how to deal with them. The first column lists the key issues that need addressing. In the second column, explain how you would complete the works.

Your tutor will want to see how you have dealt with these issues so be sure you have addressed them all.

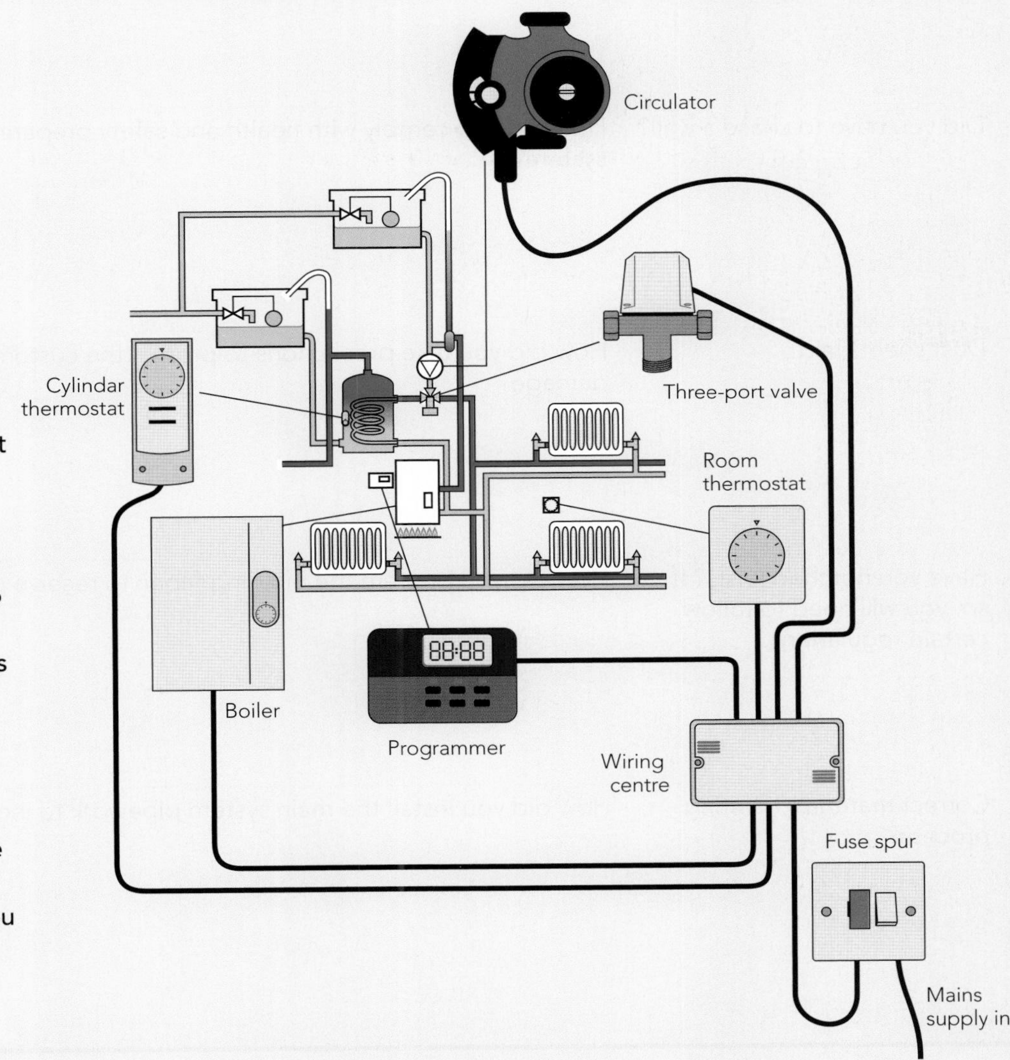

Key issues you need to mention in your write-up	Action taken
State the documents you would use to aid you with the installation. Regulations?	How did you check that all necessary plans, specifications and verbal instructions were available before starting the job?
Do you need to keep the customer informed of progress?	How did you effectively liaise with trades and the customer throughout the task?
Did you check the materials list for the correct materials and appliances needed?	Did you check that all tools required were present and safe to use and ensure that all materials were present and correct?
Did you have to chase a wall?	How did you comply with health and safety preparing and installing the system?
Dust sheets laid?	How did you take precautions to protect the customer's property from damage?
Have you notched joists? If so, you will need to follow certain regulations.	How did you prepare the building fabric to receive pipework?
Correct materials, jointing process?	How did you install the main system pipework to industry standards?

Key issues you need to mention in your write-up	Action taken
Have you connected all the correct WRAS fittings?	How did you install the cylinder and the boiler, etc. to industry standards?
Did you cap the pipework?	If required, how did you take precautions to prevent the use of unfinished installations?
Have you soldered all the fittings, tightened the DOC?	How did you carry out checks on fittings and fixtures?
How did you test the system?	How did you carry out soundness tests for the system in accordance with industry standards?
Did you carry out pressure and flow rates? How?	How did you carry out flow and pressure tests to standard?
Do you need to protect the pipework in the loft?	Did you fix insulation as required? If so, where?
Have you checked the operating procedures of all the valves?	How did you commission the system and pass over the completed job to the supervisor and customer? What information did you give?

Decommissioning heating systems

Describe what steps you would take to decommission the following systems.

1. Full drain-down of an indirect three-port system.

2. Decommission a back boiler system.

3. Decommission a radiator without draining the full system.

4. Decommission a combination system.

Your questions answered...

When replacing a boiler I noticed the water was a black colour. What can I do to clean the system to get rid of this black stuff?

Firstly, the black water and the thicker products coming out are known as sludge. This is formed by a reaction between the water inside the system and ferrous metals in the system, which produces ferrous oxide. The heavier particles will fall to the bottom of the radiators or small pipework, affecting the efficiency of the system.

When replacing a boiler the manufacturer's instructions demand that the system is cleaned during the installation process. If this is not done, the manufacturer's warranty will not be valid. The best way to do this is to use a power flushing kit. You must follow the manufacturer's instructions when using the machine to correctly clean the system. Its basic working principles are to circulate chemicals to individual radiators one at a time, flushing out the sludge.

Once this has been done and the boiler has been installed you can introduce a chemical called an inhibiter, which will keep the system clean for longer by decreasing the rate of metal oxidation.

QUICK QUIZ

1. What are central heating systems that work solely from convection known as?
2. What is the flow and return pipework to the radiators known as?
3. What is the flow and return to the cylinder called if a port valve has been used?
4. What is a bypass?
5. What is the water storage container that fills the heating system called?
6. What is SEDBUK?
7. What are radiators and fan convectors otherwise known as?
8. What does TRV stand for?
9. Why should you use a room thermostat?
10. What is a condensing boiler?
11. What do we mean by the word 'combination' in a combination boiler?

Unit 009/209

Understand and apply domestic rainwater system installation and maintenance

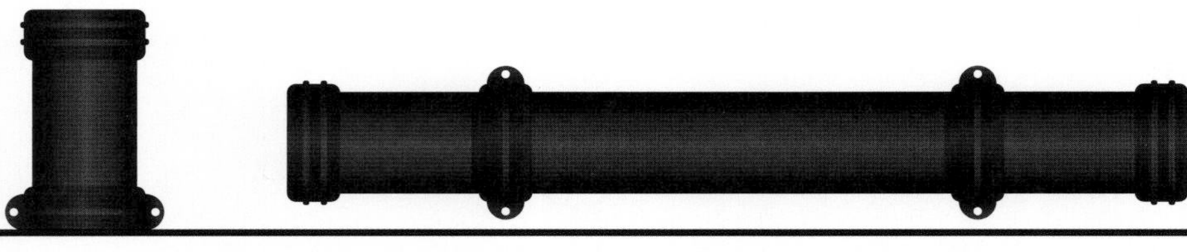

When we discuss rainwater systems we are essentially talking about the collection of rainwater that falls on to a surface and the disposal of that water. If there was no adequate rainwater system, damage to property would almost certainly occur. This damage is known as weathering.

Most guttering fitted today is made of PVC-U, which makes it light, easy and cheap to install. The most common old systems were made of cast iron or asbestos, and you may even see extruded aluminium. You will need to be careful when removing or working on these old types of guttering as they can present a risk to health.

You must follow the Building and Working at Height Regulations for this unit. See your tutor for copies.

Learning outcomes

- Know the general principles of gravity rainwater systems
- Know the layout requirements of gravity rainwater systems
- Know the site preparation techniques for gravity rainwater systems
- Be able to apply site preparation techniques for gravity rainwater systems
- Know the installation requirements of gravity rainwater systems
- Be able to install gravity rainwater systems
- Know the service and maintenance requirements of gravity rainwater systems
- Be able to service and maintain gravity rainwater systems
- Know the inspection and testing requirements of gravity rainwater systems
- Be able to inspect and test gravity rainwater systems

Key knowledge

- Rainwater systems and pipework layouts
- Fittings
- Understand installation of systems and procedures to follow for safe installation
- Maintaining and decommissioning guttering systems

Rainwater systems and pipework layouts

Types of guttering

ACTIVITY

Draw a cross-section of the THREE different types of guttering pipework you will work with:

1. Ogee

2. Square line

3. Half round

Flow rates

The table below shows the typical flow rates of the different types of guttering listed on the previous page.

Flow rates

Outlet at end of gutter run				
System	Gutter fixed level		Gutter fixed at 1:600 fall	
	Gutter flow (litres/sec)	Roof area (m^2)	Gutter flow (litres/sec)	Roof area (m^2)
Half round	0.92	44	1.17	56
Square line	1.70	82	1.18	88
Ogee	2.34	122	2.50	122
Outlet at centre of gutter run				
Half round	2.60	125	2.75	132
Square line	3.41	160	3.95	190
Ogee	6.14	294	6.25	300

Typical rainwater pipework layout system

Rainwater gravity system

On the diagram below, calculate the correct fall and clip spacing. Show your workings for both using the table above.

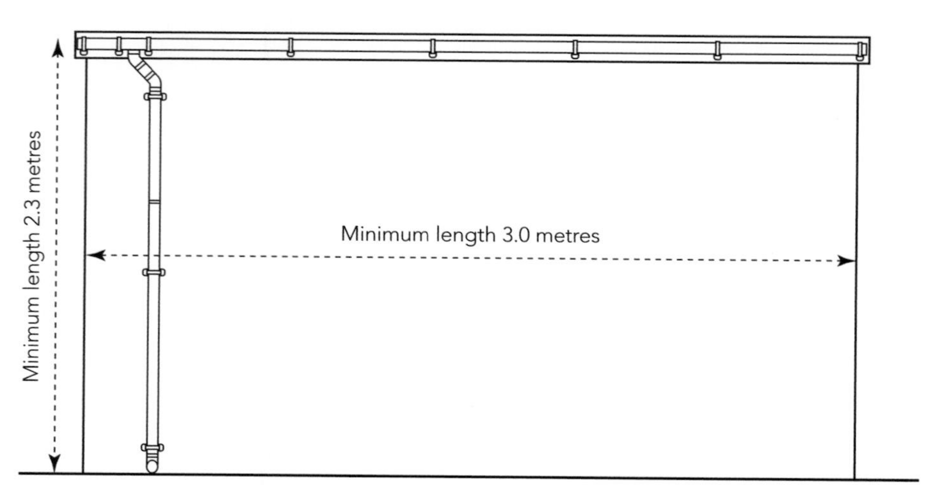

Fall and clip spacing

Fittings

Complete the diagram by naming the individual fittings shown.

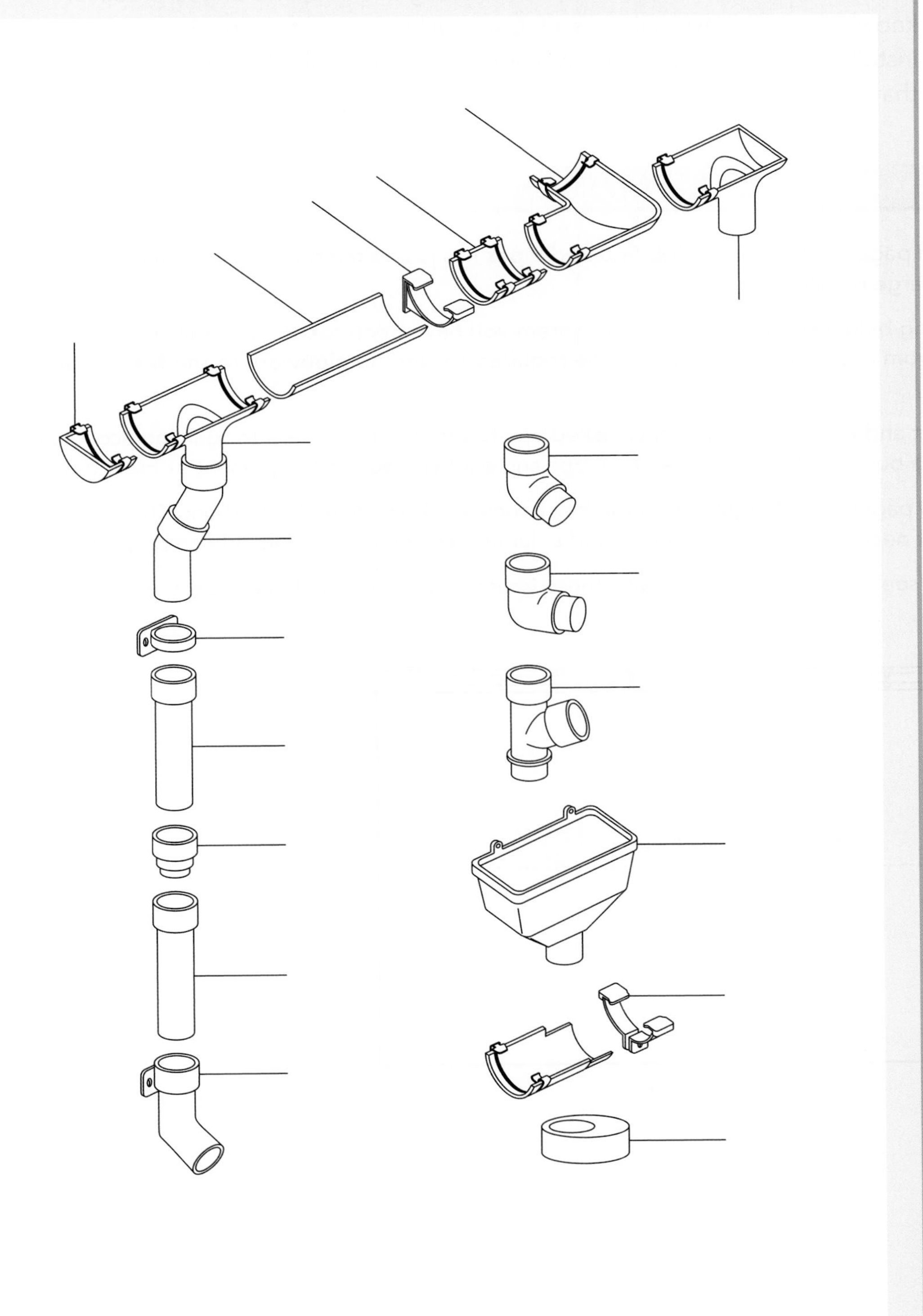

Types of guttering

Understand installation of systems and procedures to follow for safe installation

key terms

Fascia board: the board that is fitted to the frame of a roof under the tiles.

In this section you will be given a project to complete. Using all the information you have obtained from the units, including this one, and all the practical knowledge you have gained so far, you will be asked to consider a scenario for an installaton as a plumber. This will also be strongly linked to the on site assessments that you will need to complete.

ACTIVITY MANAGE MANAGE

Complete the table on pages 167–8. PlumbBob Squared Pipes has just taken on a job installing a new guttering system in a large dwelling.

You will use the drawing below showing the type of system you have been asked to install. You are installing the system from new into a gully. You will be required to clip the pipework to the **fascia board** and to a solid wall.

The customers, Bridget and Anthony Wilson, have asked you to arrange the means for gaining access in a safe way. They live on a busy road where there are shops and a lot of pedestrians present at most times.

Complete the table on pages 167–8 highlighting the key issues and how to deal with them. The first column lists the key issues that need addressing. In the second column, explain how you would complete the works.

Your tutor will want to see how you have dealt with these issues so be sure you have addressed them all.

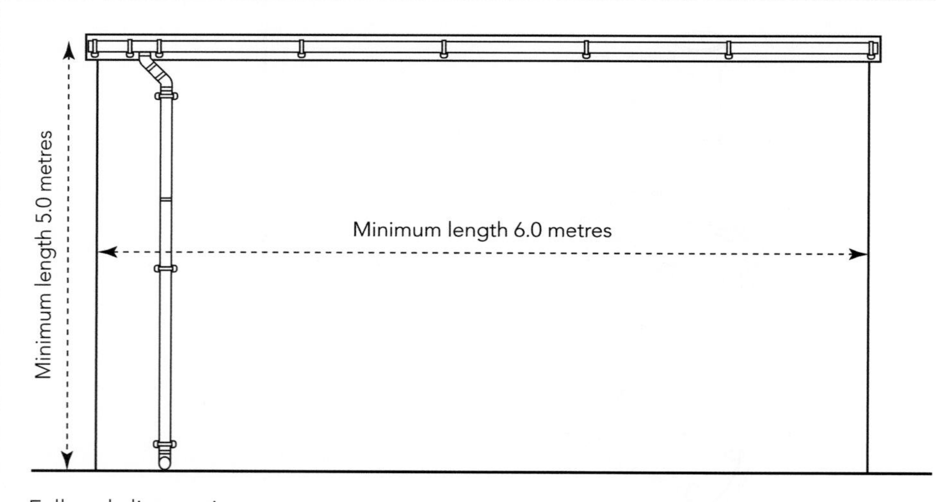

Fall and clip spacing

Key issues you need to mention in your write-up	Action taken
State the documents you would use to aid you with the installation. Regulations?	How did you check that all necessary plans, specifications and verbal instructions were available before starting the job?
Do you need to keep the customer informed of progress?	How did you effectively liaise with trades and the customer throughout the task?
Did you check the materials list for the correct materials and appliances needed?	Did you check that all tools required were present and safe to use and ensure that all materials were present and correct?
Did you have to chase a wall? Put up scaffolding? PPE?	How did you comply with health and safety preparing and installing the system?
Dust sheets laid?	What did you do to take precautions to protect the customer's property from damage?
Have you notched joists? If so, you will need to follow certain regulations.	What did you do to prepare the building fabric to receive pipework?

Key issues you need to mention in your write-up	Action taken
Correct materials, jointing process? Falls?	What is the correct procedure to install the main system pipework to industry standards?
Did you cap the pipework?	If required, how did you take precautions to prevent the use of unfinished installations?
Have you clipped every fitting correctly?	What checks did you make to fittings and fixtures?
How did you test the system?	What's the procedure for soundness testing the system in accordance with industry standards?
Did you carry out pressure and flow rates? How?	Explain how you carried out flow and pressure tests where applicable.
	How did you commission the system and pass over the completed job to the supervisor and customer? What information did you give?

Decommissioning and maintaining guttering systems

ACTIVITY

Describe what steps you would take to decommission the following systems.

1. Full decommission of an asbestos system.

2. Decommission of a cast iron system.

3. Clearing a blocked downpipe.

Your questions answered...

The guttering I have fitted keeps on popping off the stop ends in hot weather. Why?

The simple answer is that you have not done the installation correctly. If you get a fitting from a guttering system and have a look inside you will see a raised line, known as the fitting line. The length of guttering you fit should not hang over that line as it is for the natural expansion of the plastic due to its coefficient value. You will find more information about this in unit 204.

You may have fittings without this line so you will need to work out the expansion yourself. You should leave enough room from the end of fittings for the adjustment. Below is an example:

length (m) × temperature rise (°C) × coefficient value of material = the total the material will rise

example: 5 m of plastic × 8 °C rise in temperature × 0.00018 (value) = 0.0072 m or 7.2 mm

Remember: 1 metre = 100 cm = 1,000 mm

QUICK QUIZ

True or false?

1. Downpipes have shoes.
2. The fall for guttering systems is 1:6,000.
3. You can get running inlets.
4. LCS is a common system.
5. There are TWO different types of stop end available.
6. When working at heights it is OK to use scaffolding.
7. Clip spacing should be a minimum of 2 m.
8. PVC is heavy and hard to cut.
9. You can remove asbestos and throw it in the bin.
10. The THREE main types of guttering are ogee, half round and rectangle.

Unit 010/210

Understand and apply domestic above ground drainage system installation and maintenance techniques

One of the greatest inventions, which has saved countless human lives, is the humble sewer. Created to clean the streets of waste, this maze of underground pipework removes all your waste products and carries them away to be processed safely. Along with the WC, the basin, the bath and the shower, the simple sewer is an aid to comfort and hygiene, but also helps to prevent deadly diseases, which thrive in insanitary conditions.

By now you should be realising that without plumbers, heating engineers and air conditioning personnel the world would be a very different, dirty and uncomfortable place to live. As an apprentice, you may not be in direct contact with below ground drainage, but you will certainly have to work with above ground drainage systems. Understanding the processes and principles will help you to carry on the fine tradition of preventing people from becoming ill.

BS 8000 Part 13 covers the Code of Practice for above ground drainage and sanitary appliances, while BS 8313 gives guidance for creating chases, holes and ducts for pipework. You should also look back at the Water Regulations for waste of water and the prevention of backflow and Part H of the Building Regulations.

Learning outcomes

- Know the uses of sanitary appliances and their operating principles
- Know the types of sanitary pipework system and system layout requirements
- Know the site preparation techniques for sanitary appliances and connecting pipework systems
- Be able to apply site preparation techniques for sanitary appliances and connecting pipework systems
- Know the installation requirements of sanitary appliances and connecting pipework systems
- Be able to install sanitary appliances and connecting pipework systems
- Know the service and maintenance requirements of sanitary appliances and connecting pipework systems
- Be able to service and maintain sanitary appliances and connecting pipework systems
- Know the decommissioning requirements of sanitary appliances and connecting pipework systems
- Be able to decommission sanitary appliances and connecting pipework systems
- Know the inspection and soundness testing requirements of sanitary appliances and connecting pipework systems
- Be able to inspect and soundness test sanitary appliances and connecting pipework systems

Key knowledge

- Above ground drainage systems and pipework layouts
- Fittings and traps and common problems
- Sanitary appliances
- Understand installation of systems and procedures to follow for safe installation
- Maintaining and decommissioning guttering systems

Above ground drainage systems and pipework layouts

Principles of appliances for discharging waste water

ACTIVITY

Cut out and stick in pictures from magazines etc. of the appliances listed below or draw a simple diagram of each. Write a simple sentence about each appliance, explaining its use and how you think the waste water is removed. Is there anything notable about the design – for instance a dual flush on a WC?

Appliance	Your photo or diagram	Sentence of the basic working principle/design
WC		
Basin		

Bath		
Kitchen sink		
Macerator system		
Shower		
Bidet		

Primary ventilated stacks

On the diagram below, write down the minimum sized pipework for all pipework present.

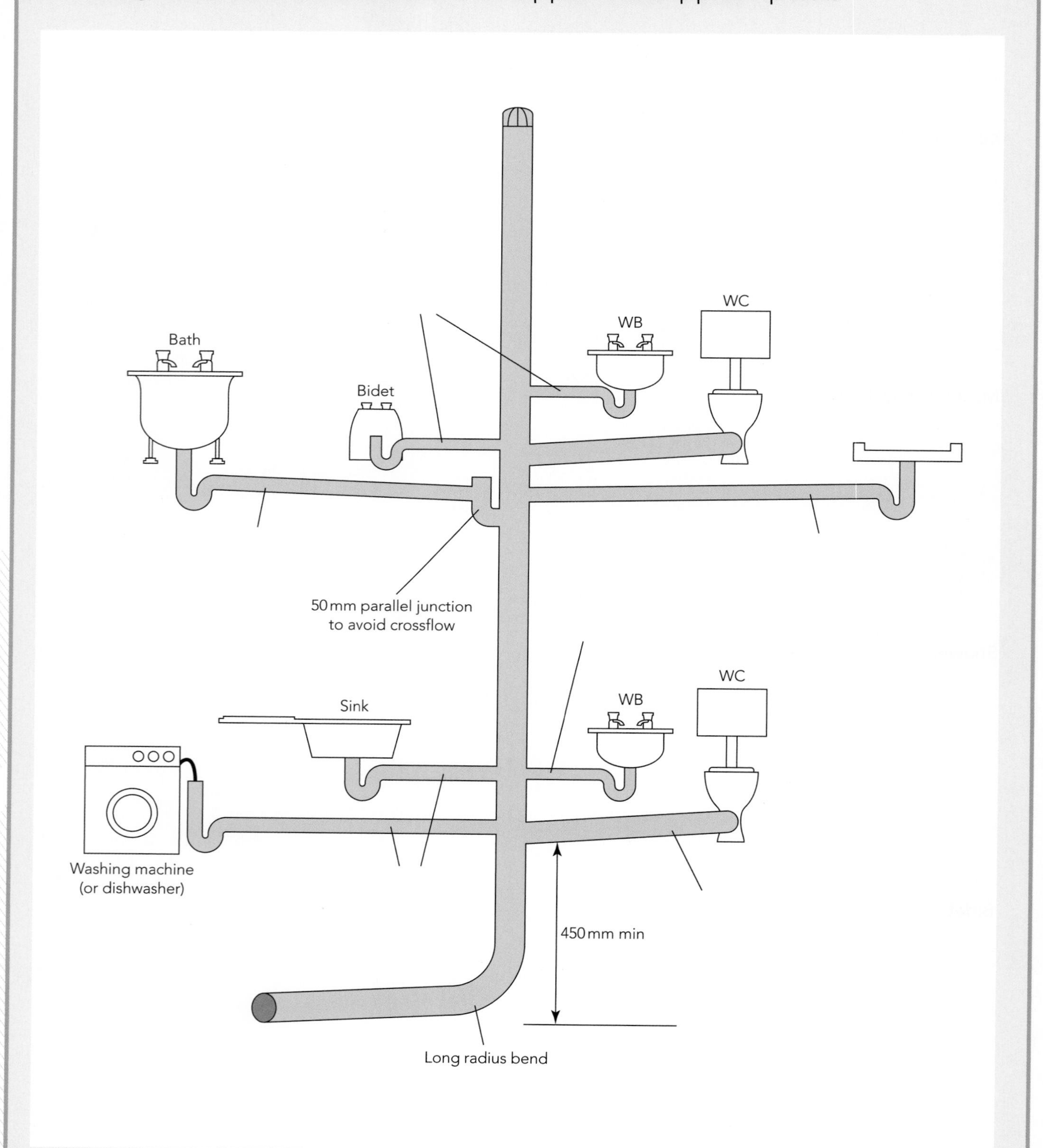

External two pipe ventilated stacks

ACTIVITY

External two pipe ventilated stacks

Look at the diagram below. Write a short statement about where the foul smells are coming from and how they might be prevented.

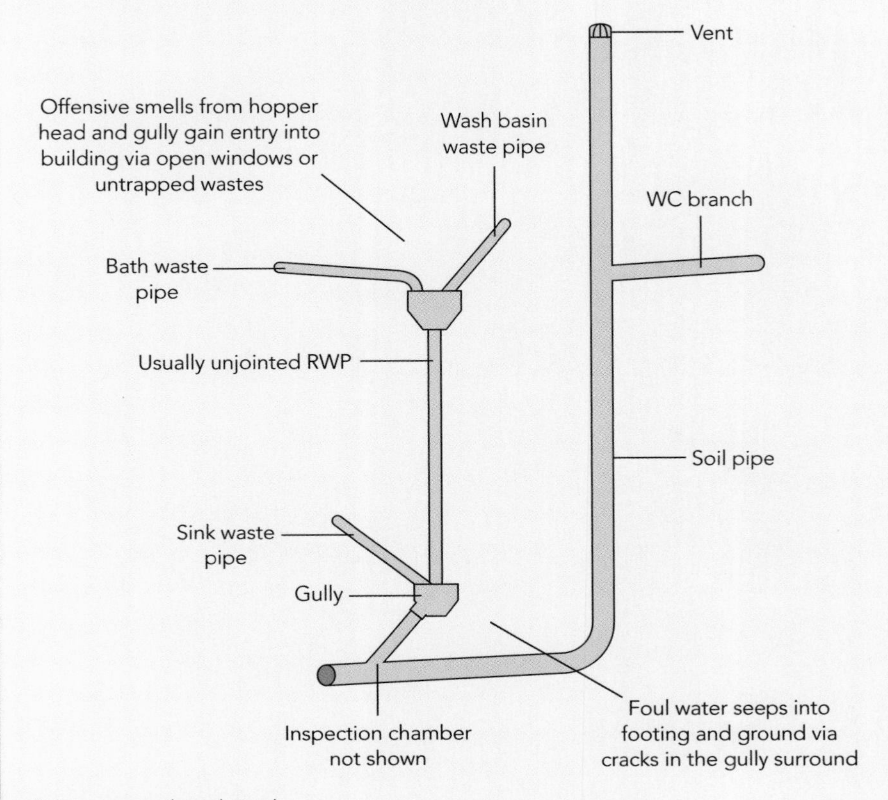

Primary ventilated stack

key terms

External: fitted outside the dwelling. Conversely, *internal* means fitted inside.

key terms

Stack: the soil waste system pipework.

Discharge branch and secondary ventilated stacks

ACTIVITY

In the space below, draw diagrams showing the main features of discharge branch and ventilated stacks. Use the internet and other literature to help you.

Limitations of use

All pipework has maximum lengths that can be used. Waste pipework must also be fitted with a slight slope. This is done for various reasons.

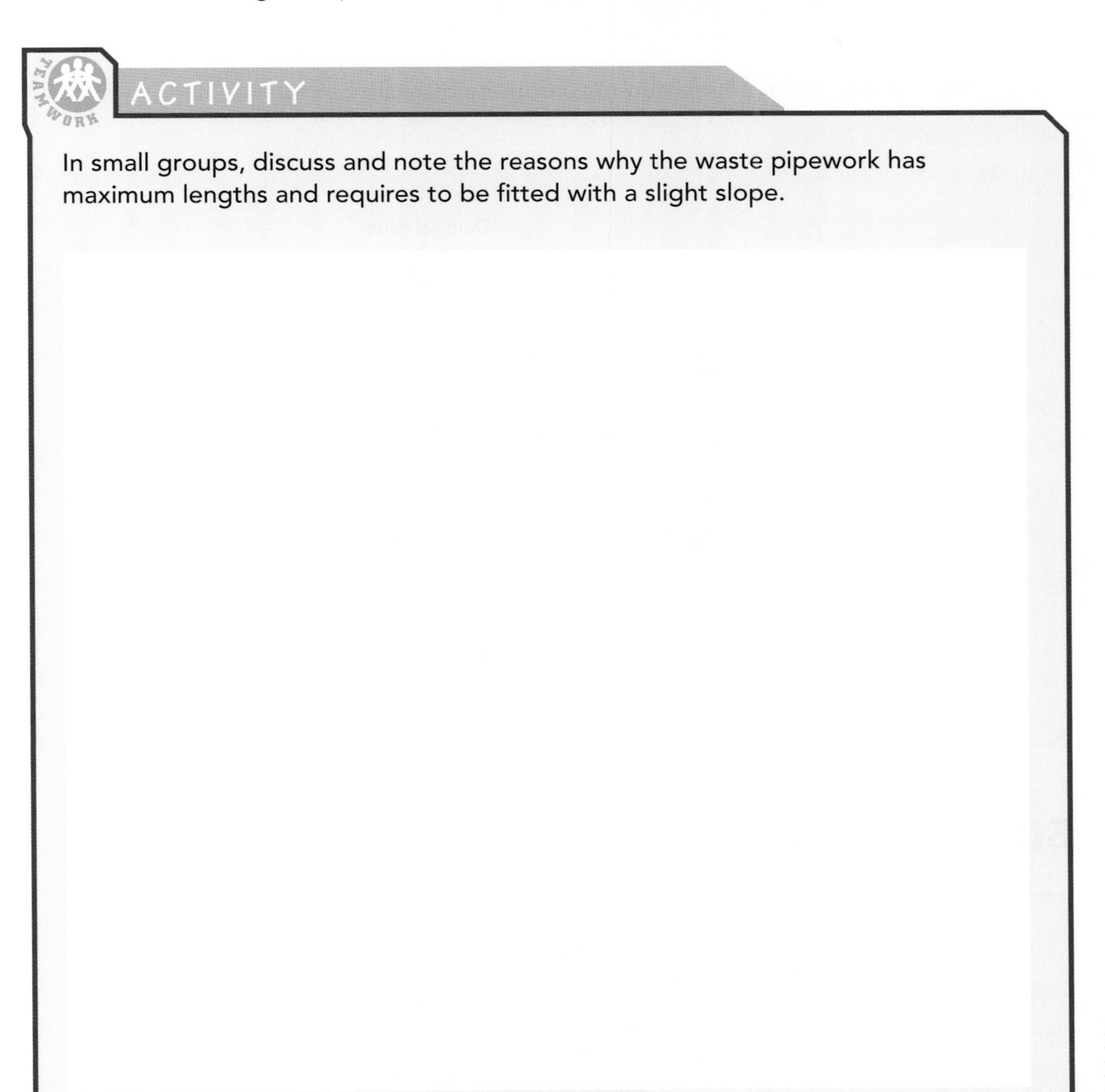

In small groups, discuss and note the reasons why the waste pipework has maximum lengths and requires to be fitted with a slight slope.

ACTIVITY

Fill in the missing data in the table below.

Pipesize	Primary use of pipework	Maximum length	Horizontal clip spacing	Vertical clip spacing	Slope
32 mm/1¼ inch					
40 mm/1½ inch					
110 mm/4 inch					

Ventilation requirements

ACTIVITY

Fill in the missing minimum measurements on the diagram below.

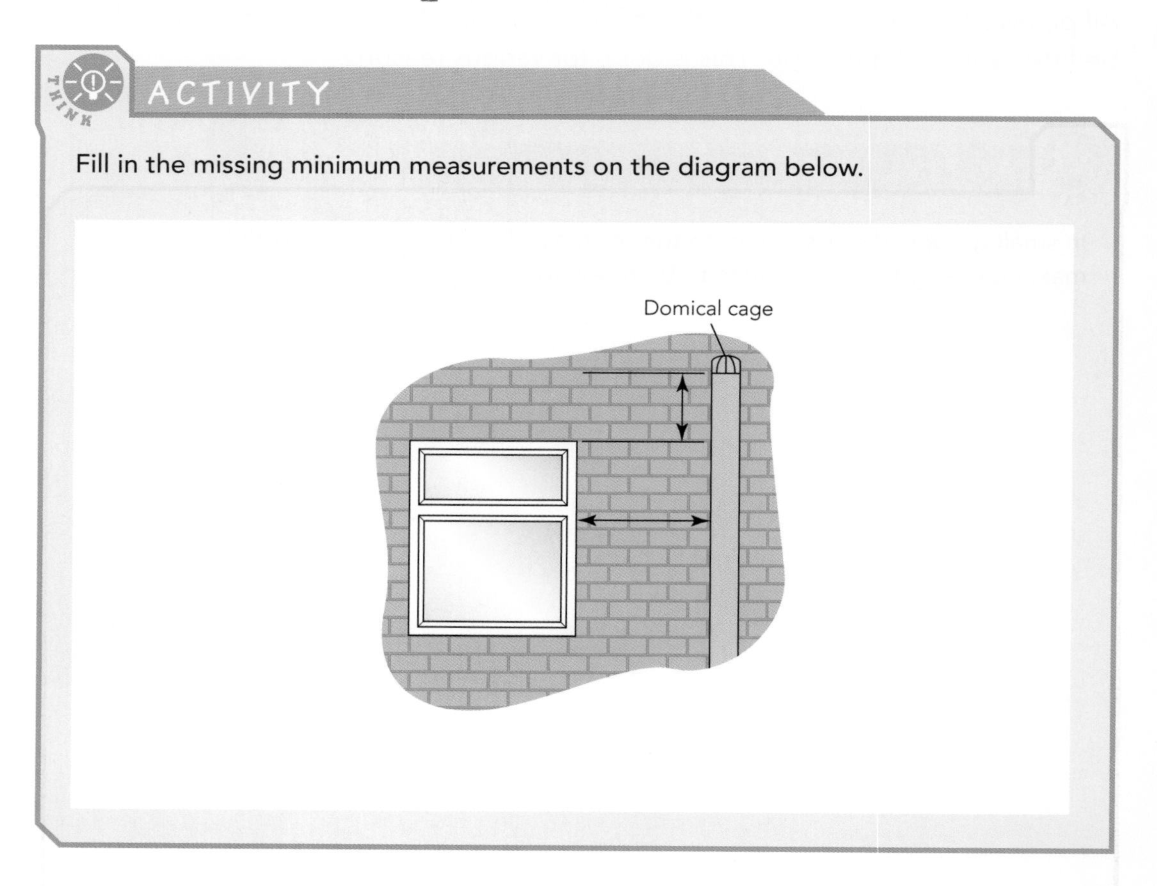

Stub stacks

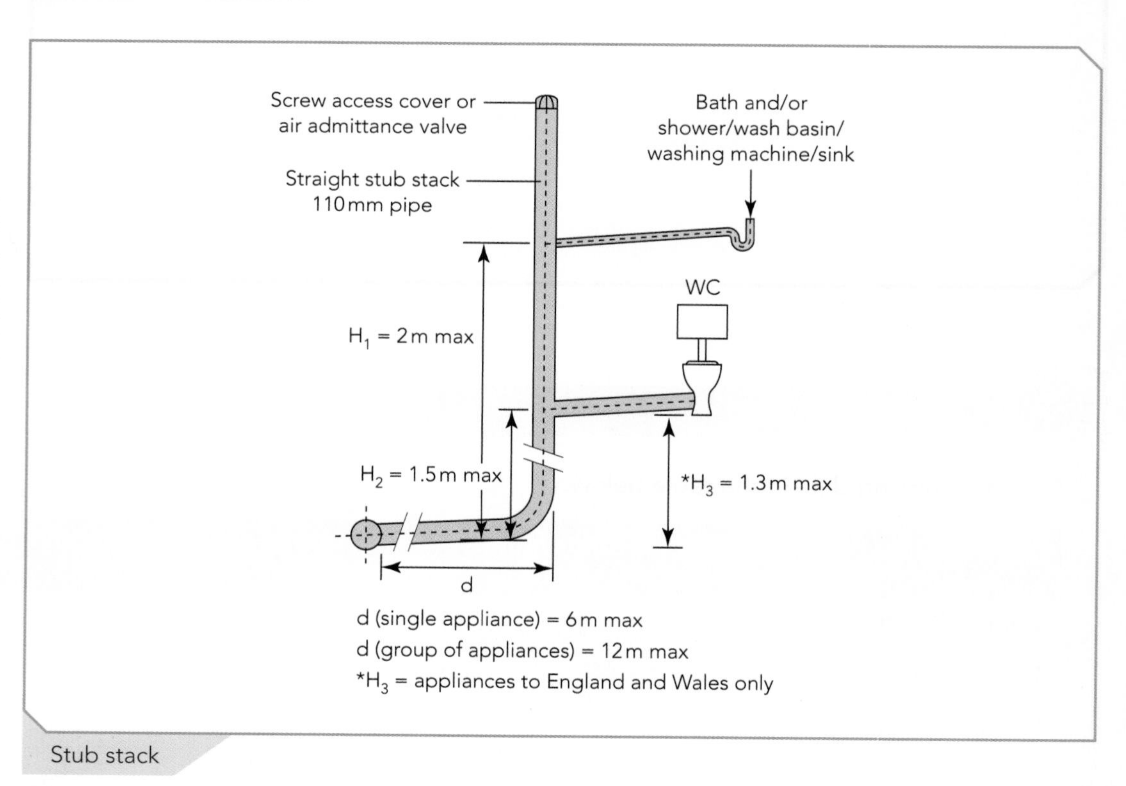

Stub stack

ACTIVITY

In small groups, discuss the circumstances where stub stacks can be used. Write up your conclusions.

ACTIVITY

Which of these statements are true in relation to venting stub stacks?

1. A stub stack can be **vented** using a 'bird's nest' at whatever point you wish.
2. An AAV must be used but only outside and works with positive pressure.
3. An AAV must only be used inside and works when negative pressure has occurred.

key terms

Vented: ventilated.

Fittings and traps and common problems

You may have noticed that the pipework you work with has an array of different fittings and, depending on where you are in the country, these can be called many different names. It can be quite confusing.

When working with wastes you will work with pipework from 110mm down to 32mm. Some people will use imperial units to describe these sizes – that is inches. When talking to your tutor at college it shouldn't matter what you use as they will understand what you mean, but please be aware that for the test you will need to understand both.

ACTIVITY

Draw a line to connect each photo with its correct statement. Name the fitting described.

These fittings can be either solvent or pushfit but are generally used with the pushfit system.

Name: ______________________

You must wear gloves when making these fittings.

Name: ______________________

Not really preferred by plumbers. When fitting it is important that the 'O' ring does not get dislodged.

Name: ______________________

These fittings allow you to connect on to multiple types of pipework to suit your needs.

Name: ______________________

Waste traps

ACTIVITY

In pairs, discuss what you think waste traps are and why you think they are needed. What is their primary purpose?

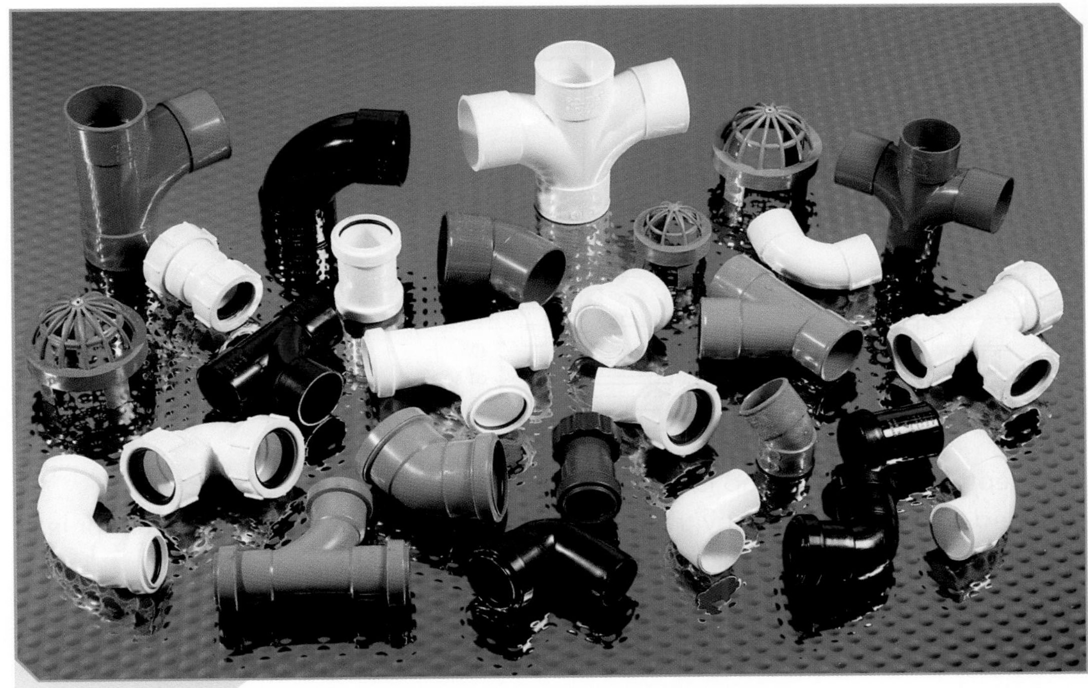

Waste traps

When you go home look at the traps used in your dwelling. Note the types and where they are used. Draw diagrams.

In groups, discuss the traps you have found. If you have an unusual looking trap, highlight this with the group and discuss with your tutor why you think it has been used.

Places to look:

- kitchen sink
- washing machine
- basins
- baths/showers.

Note: you might not always be able to easily locate or gain access to some traps. Only examine the ones that are straightforward to find.

Waste seals

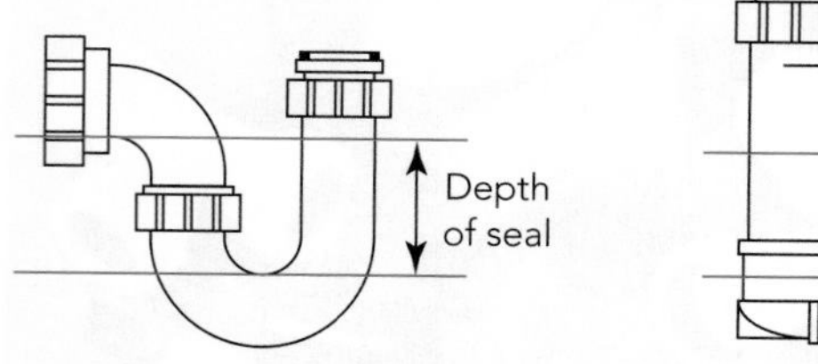

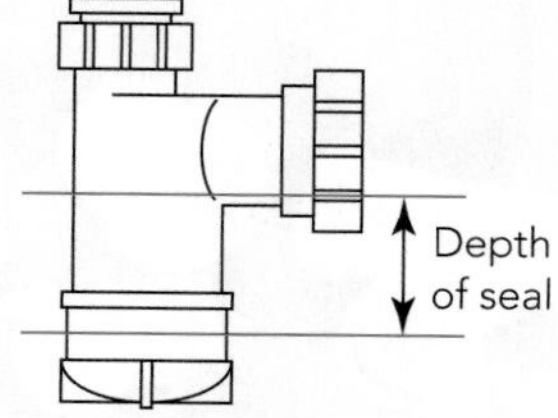

The depth of water in a trap is important and BS EN 12056 Part 2 must be followed. This states that 'any trap connected to a discharge pipe of 50 mm or less discharging into a main stack should have a seal of 75 mm'.

A trap diameter of 50 mm or above must have a trap seal of 50 mm. When the water is discharged into a gully or hopper head, a minimum seal of 38 mm is permitted.

ACTIVITY

Fill in the missing measurements. You can use the internet and other literature to help you.

Appliance	Trap size	Seal required in mm
WC		
Basin		
Bath		
Kitchen sink		
Washing machine		
Shower		
Bidet		

Loss of seal in the trap

Pipework design and natural causes can take away the seal of the trap, which will lead to smells and unhappy company bosses and customers! Here, we will look at three scenarios.

ACTIVITY

Self-siphonage

Study the drawing above and write down what you think is happening at the THREE points. You can use the internet to help you.

1.

2.

3.

Induced siphonage

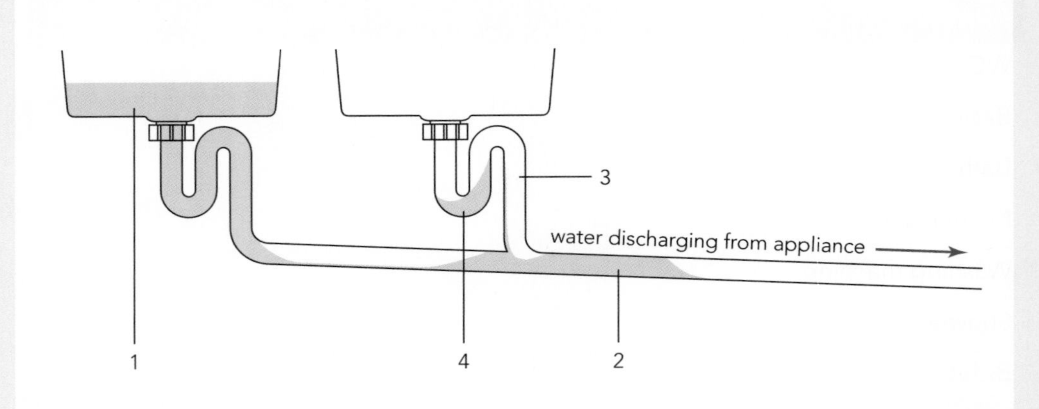

Study the drawing above and write down what you think is happening at the FOUR points. You can use the internet to help you.

1.

2.

3.

4.

Compression

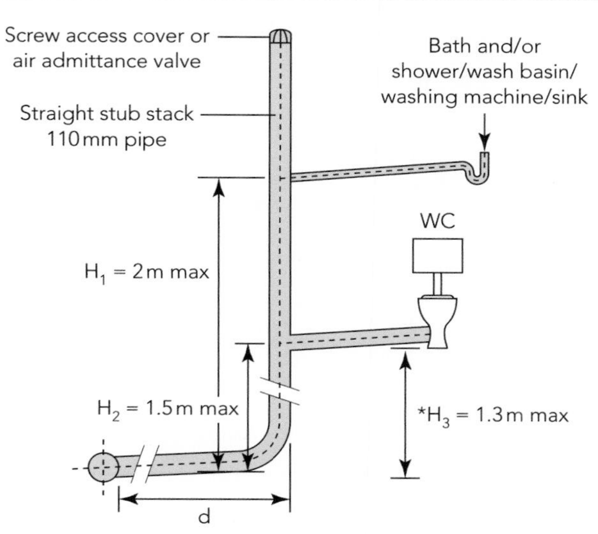

Explain what is meant by compression if the incorrect bend is fitted instead of the correct sweeping bend that is required at the bottom of the diagram.

In groups of three, discuss how you correct the siphonage problems shown in the three previous activities. Write down your conclusions.

Self-siphonage

Induced siphonage

Compression

Natural phenomena

Explain the causes of the following phenomena. How might they adversely affect a plumbing installation? What could you do to prevent them?

Action	Cause	Effect	Prevention
Evaporation			
Momentum			
Capillary action			
Wavering			

Sanitary appliances

WC pans and cisterns

There are four different types of WC manufactured today:

- close-coupled
- high level
- low level
- concealed.

Water Regulations state that a maximum level of 6 litres should be used for flushes. New style dual-flush systems are the normal types you will be fitting in modern installations and these will come in 4 litre and 6 litre full flush specifications. Along with this, you can also have different types of pan – a simple wash down pan will be the normal configuration you will see, but you can also get single and double trap siphonic pans. You should understand what siphonic action is. If you're still unsure, go through this unit or use the internet to find diagrams of the two different types of siphonage.

ACTIVITY

Older siphon type configuration

Label the components of the cistern shown below.

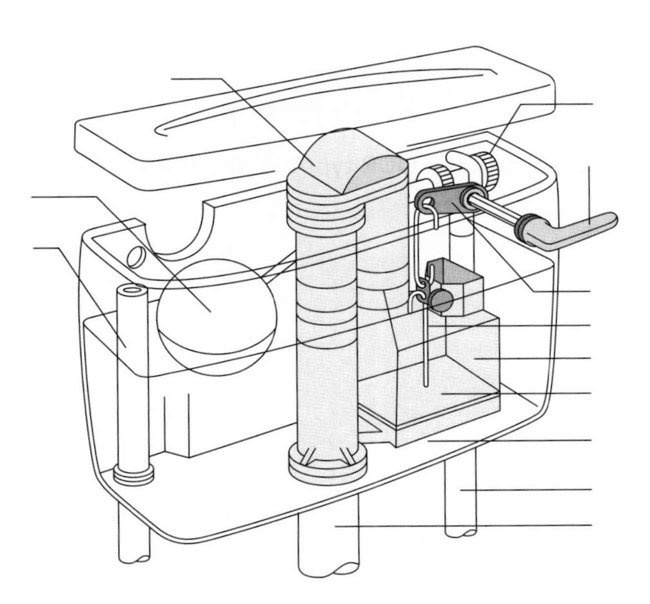

ACTIVITY

New dual-flush system configuration

Label the components of the cistern shown below.

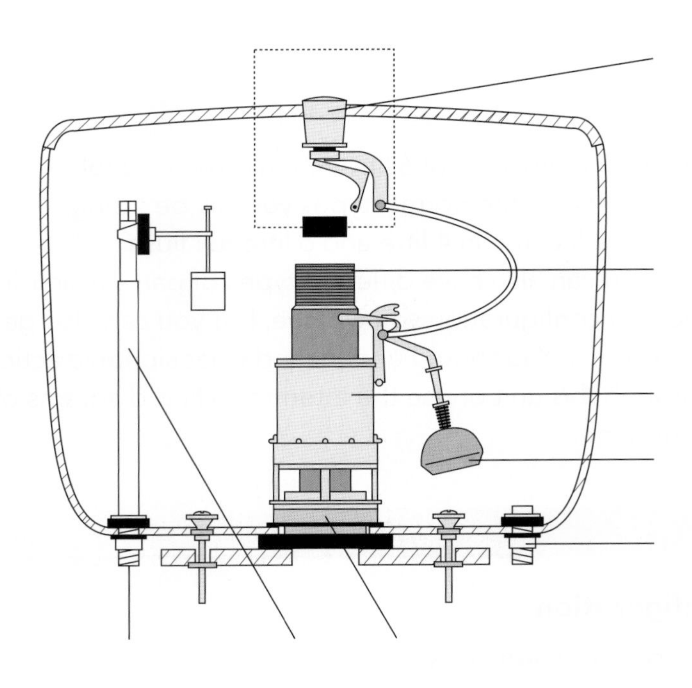

ACTIVITY

In groups, discuss the differences between the FOUR types of WC listed on the previous page. Make notes on your findings.

Washbasins

There are also four main types of washbasin:

- pedestal
- wall hung
- counter top
- under counter top, including semi.

These types also come with various tap configurations, such as one hole, two hole and three hole. There are also different overflow types, namely integral or not. Whatever configuration you have, you must follow the manufacturer's instructions when fitting taps and waste systems to the appliance.

ACTIVITY

Label the diagram below for a normal configuration of a waste system to the basin.

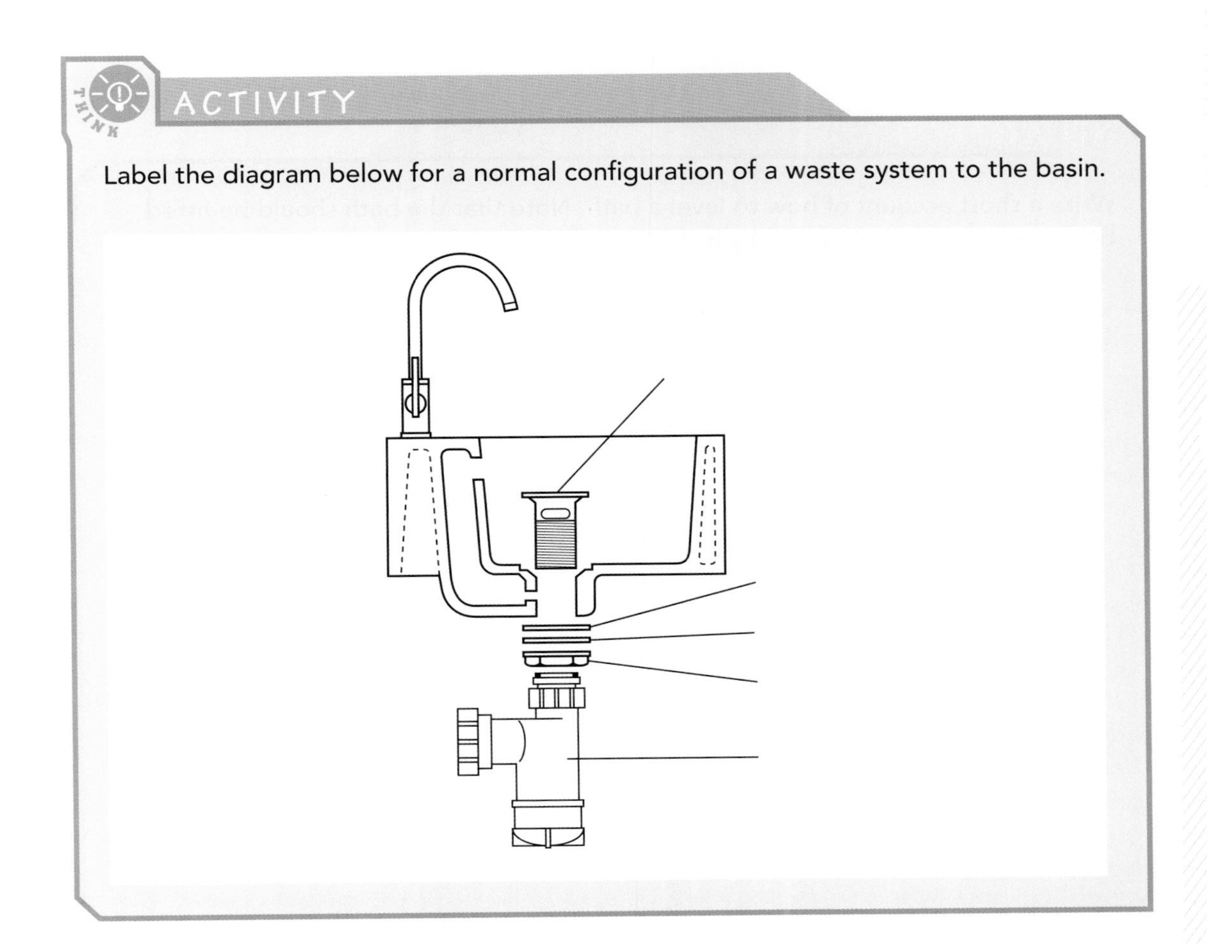

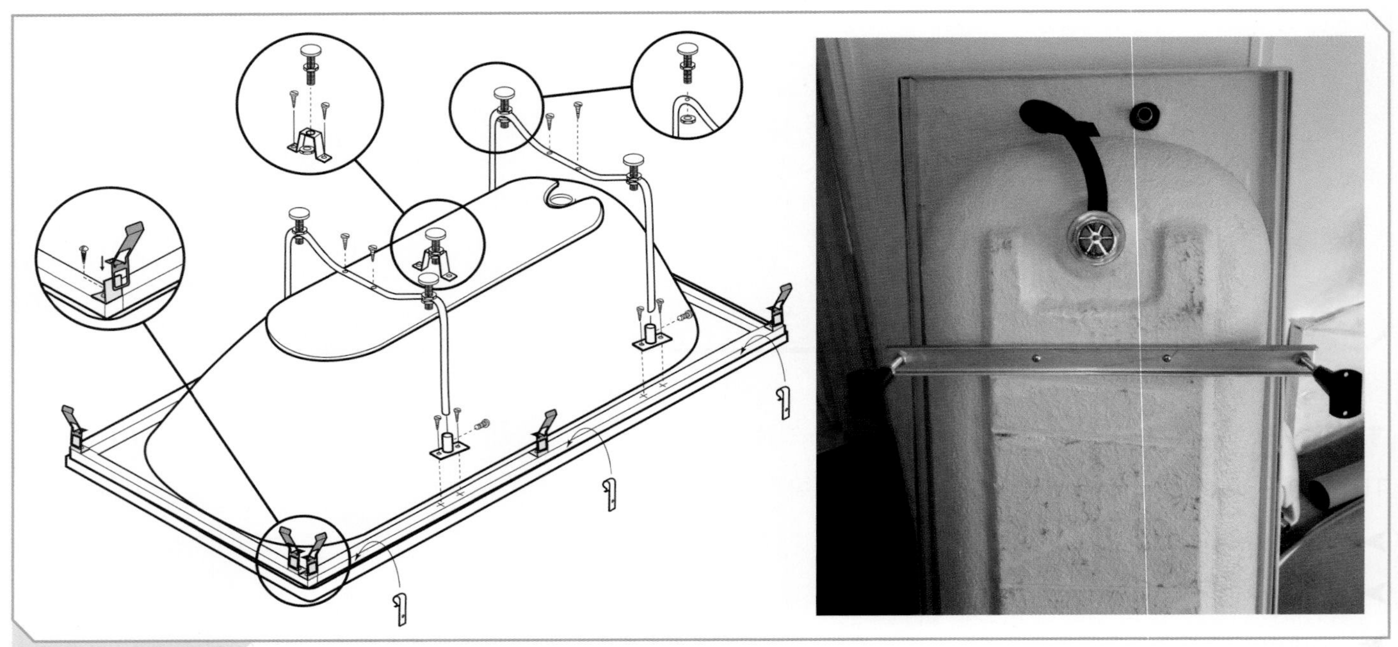

Bath parts

Baths

Baths are made out of two main materials: mild steel or plastic. They do not come fitted and some will require you to cut the tap holes to suit. Ask your boss to show you this as it can be a tricky process and you will learn best by watching first and then doing.

ACTIVITY

Write a short account of how to level a bath. Note that the bath should be fitted between 500 and 550 mm – why is this?

Shower trays

The two main materials for shower trays are plastic and stone resin.

Shower parts

ACTIVITY

THINK

In small groups, discuss how you would support a stone resin tray if it couldn't be placed on the floor. Make notes on your discussion.

Bidets

Bidets are designed for personal hygiene use. They come in two types: over the rim and ascending spray. Because of the waste products that are present, the rules under the Water Regulations differ.

ACTIVITY

The bidet in the picture below is an above rim supply – there is a natural air break between the water outlet and the pan. This conforms to the Water Regulations.

Using the internet and the regulations to help you, discuss the following in pairs:

- What are the limitations for ascending spray types?
- What do you need to do to comply with the regulations?

Write up your findings.

Urinals

Urinals are fitted in non-domestic buildings. They can be steel or slab urinals or individual bowls. Urinals do not have the best reputation for saving water. However, new technology means that some urinals now use hardly any water at all.

THINK ACTIVITY

Study the diagram below. Describe what happens at point 1 to make the siphon automatic.

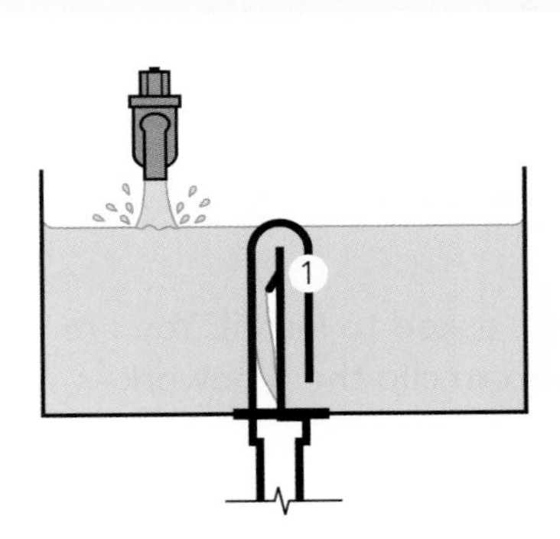

List the different valves or pieces of technology that help with the prevention of waste of water. You can carry out some research using the internet.

Understand installation of systems and procedures to follow for safe installation

In this section you will be given a project to complete. Using all the information you have obtained from the units, including this one, and all the practical knowledge you have gained so far you will be asked to consider a scenario for an installation as a plumber for an AGDS (above ground discharge system). This will also be strongly linked to the on site assessments that you will need to complete.

ACTIVITY

Complete the table on pages 195–7. PlumbBob Squared Pipes has just taken on a job installing a new AGDS in a large dwelling.

You will use the drawing on page 195 showing the type of system you have been asked to install. You are installing the system from new to the underground drainage. You will be required to clip the pipework correctly.

The customers, John and Joan Smith, have asked you to install all the pipework to the outside of the dwelling. There are no issues regarding the walls and the job is very straightforward. They will be present when you start and they wish you to keep them informed of progress. The customers are very house proud and Mrs Smith is very keen on cleanliness. You must not upset Mrs Smith.

The sanitaryware is also to be installed and you will use the specification verbally given by the Smiths. Again, there is no issue for the installation of the sanitaryware. The tiler is due in after you to complete the splashbacks.

Please make any notes you need before you start.

Halfway through the install, Mrs Smith has requested you change the bidet and the WC around from the original positions agreed. She has also noticed that the wall has a scuff where something has been dragged along it. Please ensure you mention how you would deal with these problems.

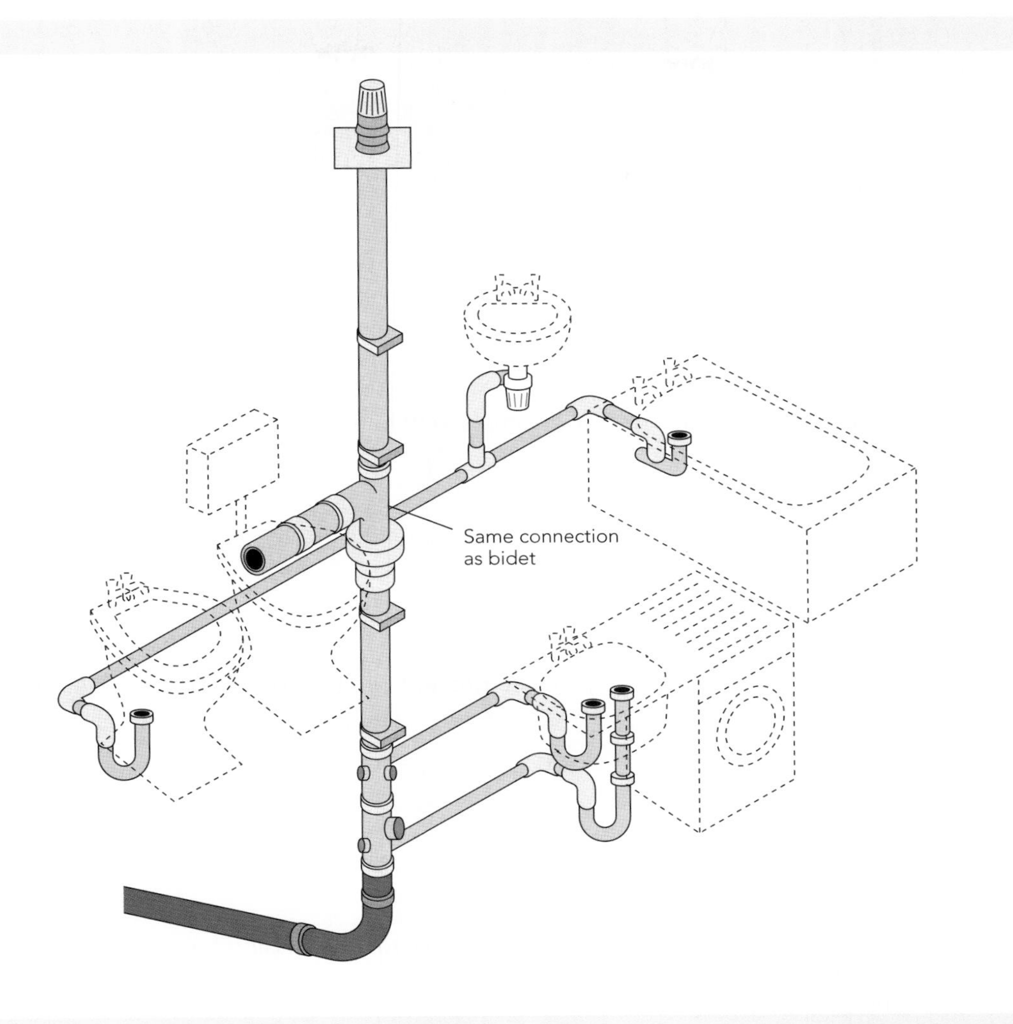

Complete the table on pages 195–7 highlighting the key issues and how to deal with them. The first column lists the key issues that need addressing. In the second column, explain how you would complete the works.

Your tutor will want to see how you have dealt with these issues, so be sure you have them all.

Key issues you need to mention in your write-up	Action taken
State the documents you would use to aid you with the installation. Regulations?	How did you check that all necessary plans, specifications and verbal instructions were available before starting the job?
Do you need to keep the customer informed of progress?	How did you effectively liaise with trades and the customer throughout the task?

Key issues you need to mention in your write-up	Action taken
Did you check the materials list for the correct materials and appliances needed?	Did you check that all tools required were present and safe to use and ensure that all materials were present and correct?
Did you have to chase a wall? PPE?	How did you comply with health and safety preparing and installing the system?
Dust sheets laid?	How did you take precautions to protect the customer's property from damage?
Have you notched joists? If so, you will need to follow certain regulations.	What did you do to prepare the building fabric to receive pipework?
Correct materials, jointing process?	How did you install the main system pipework to industry standards?
Have you connected all the correct WRAS fittings?	How did you install the soil and waste pipework to industry standards? Include the appliances.

Key issues you need to mention in your write-up	Action taken
Did you cap the pipework?	If required, how did you take precautions to prevent the use of unfinished installations?
Have you soldered all the fittings and tightened the DOC?	Did you carry out checks on fitting and fixtures? If so, what?
How did you test the system?	Did you carry out soundness test to the system in accordance with industry standards?
	Would you carry out performance tests? How?
Have you checked the operating procedures of all the valves?	How did you commission the system and pass over the completed job to the supervisor and customer? What information did you give?

Decommissioning and maintaining guttering systems

ACTIVITY

1. Describe what steps you would take to decommission the following systems:

   a. Full decommission of a cast iron system.

   b. Partial decommission when removing the WC.

2. What would you have to provide if you were required to fully decommission WCs and other washing facilities for a considerable time?

Your questions answered...

I need to run the waste pipework further than permitted and I'm going to be using the solvent weld for a few hours, maybe five hours in total. What is the best way around this?

Firstly, when using solvent weld please look at the data sheet on the tub, which will state how long you are permitted to use it before you need a break. This is all down to the fumes given off. If you are in a well-ventilated area, this should not be a cause for concern, but if you are in a confined space you should stick strictly to the guidelines. Explain to your peers what you are doing and, if you start feeling funny, stop and go and see your boss.

On the second point, if you have gone over the maximum run of pipework permitted, you must increase the size to the next one up. If you have connected two appliances to one pipework run, I would advise on fitting an anti vac trap to stop any possible siphonic action that may occur.

QUICK QUIZ

1. What is the most common soil pipe system you will have to install?
2. If you are using a hopper, what type of system are you working with?
3. You can get copper waste pipework. True or false?
4. The max length for 32 mm waste is 2.7 m. True or false?
5. You can use solvent weld on pushfit fittings. True or false?
6. You can fit a cistern which takes 7 litres. True or false?
7. Urinal cisterns can operate using a solenoid valve. True or false?
8. What do we mean by capillary action in AGDS?
9. What are basins with an inbuilt overflow called?
10. What type of trap should be used to reduce gurgling noises in systems?

Glossary

Anode: Positively charged electrode.

Best practice: A statement describing the correct way to conduct a process.

Bore: Internal hole in a tube of a given size for something to travel in.

British Standards: Set standards of quality for goods and services, recognised by the award of a BS number. If something does not have a BS number, then it hasn't been approved.

BS EN: British Standard and European number. British Standards ensure the standards of quality and the dimensions are constant and correct. Where an EN has been input, this means the standard has been met in Europe.

Capillary action: The tendency of a liquid in a small space or absorbent material to rise or fall as a result of surface tension.

Cathode: Negatively charged electrode.

Chase: Channel created to run pipework in a wall or a floor to hide that pipework from the owner's view.

Cistern: A receptacle for holding water or other liquid.

Coefficient: A physical property that is constant under specified conditions.

Coil: Heated copper heating element in a cylinder fitted in circular fashion so the heat transfer can be distributed evenly throughout the cylinder.

Combustible: Capable of catching fire and burning.

Commissioning: The process of finishing a job, ensuring everything works and getting the system to operate at its optimum for the client.

Communication: Describes the various ways information is passed from one person to another, or to a group.

Competent person: Someone who has been trained and certified to carry out the work needed.

Complaint: An expression of dissatisfaction with something.

Conservation: Preservation of a commodity, such as oil or water.

Constant: Permanent or unchanging. Often applied to a mathematical value that does not change.

Contamination: The mixing of substances of which one could cause harm.

Contractors: Persons who are on site but go to and from different sites as they are not 'employed' by a company but work to a contract.

Convection: The movement of molecules within fluids (i.e. liquids, gases). It is also known as a mode of heat transfer.

Crimped: Fitting that is compressed on to a pipe using no other substance or specialist parts, such as olives.

CSCS: Construction Skills Certification Scheme.

CWSC: Cold water storage cistern.

Density: The degree of compactness of the molecules of a substance.

Dissimilar: Not alike; not similar.

Diversity: when used as a contrast or addition to equality, it is about recognising individual as well as group differences, treating people as individuals, and placing positive value on diversity in the community and in the workforce.

Earthing: Transferring of any leak of electricity or charge direct to the ground for safety.

Efficiency: Ability to produce a desired effect with minimum effort.

Electricity: The flow of electrons through a conductor.

Electromotive series: Metals ranked in order of their ability to corrode other metals by a battery effect.

Emissions: Pollution or waste energy released to the atmosphere.

End feed: The most commonly used type of capillary fitting used with solder.

Equality: Creating a fairer society, where everyone can participate and has the opportunity to fulfil their potential.

Ergonomic approach: Taking an educated approach to a situation; the word actually translates to mean 'science of work'.

External: Fitted outside the dwelling. Conversely, *internal* means fitted inside.

Fascia board: The board that is fitted to the frame of a roof under the tiles.

Ferrous: Containing iron.

Flow and return from the boiler: These are called the primaries. The flow and return to the cylinder will now become the secondaries.

Flow rate: The volume of flow per unit of time. We use litres per minute in the mechanical sector.

Fortec: A brand name of a combination cylinder where the CWSC sits directly on top of the cylinder to save space. You will hear the term Fortec more than the term combination cylinder.

Full bore: A fitting or valve that has the same diameter as the pipework it is fitted to, so the flow is not restricted.

Good practice: A term used to describe work that is not regulated but will assist with saving time, effort and resources in the future.

Gravity system: A system that works without the requirement of pumps, using only gravity and atmospheric forces.

Hazard: A situation that may be dangerous and has the potential to cause harm.

Horizontal: Going in a straight line across from side to side.

Hot water cylinder: Vessel that stores hot water for future use.

Insulation: Material used to prevent or reduce the loss of energy from a surface.

Legislation: The process for making new laws and the laws themselves.

LPG: Liquid petroleum gas.

Manipulative: Fitted to allow a small amount of movement due to its use or location.

Masonry: Stonework, such as brick, concrete and building stone.

Mass: Not to be confused with weight, mass refers to the quantity of matter a substance contains.

Mechanics: The branch of physics concerned with the working parts of a machine.

Meter (water): Device that records the amount of water used.

Minutes: Records of a meeting. A form of tracking who said what and what was said in a meeting and a record of action agreed.

Operatives: All persons who are working on site.

Optimum: The best possible conditions for a system to operate at its full potential.

Palatable (water): Water that can be drunk and tastes pleasant.

PAT: Portable appliance testing; a process in which electrical appliances are routinely checked for safety.

Patina: A film of oxide that forms on some metals.

Perceived: How somebody will view you and your personality by the way you interact with them.

Permit: Documentation issued for when you need to be given the permission to carry out an activity.

Personal protective equipment (PPE): All the equipment you use to keep yourself safe from having an accident.

Personnel: Employees taken on by an organisation to carry out the works required.

pH: Power of hydrogen – a measurement of the acidity or alkalinity of fluids.

Pi: A number that represents the ratio of the circumference of any circle to its diameter. The value of pi is 22/7 or 3.142 and this number is used in calculations to find out the dimensions of circles, spheres and cylinders.

Plug: Device used to aid in the making of a strong fixing, such as a plastic sheath around the screw that fits into drill holes and grips the masonry by means of expansion.

Potable (water): Water that is safe to drink. Water doesn't have to be palatable to be potable.

Precipitation: A deposit on the earth of hail, mist, rain, sleet or snow; also, the quantity of water deposited.

Primary flow and returns: The pipework from the boiler to the cylinder.

Products of combustion: Unburnt and created gases that have been produced from burning fuels.

PTFE: Polytetrafluoroethylene is a thin plastic, which comes on a roll and is used in the jointing process.

Pushfit: Fittings used for high-speed installations that can be taken apart again if needed. They use a series of grips and 'O' rings to create the seal.

R numbers: Measure the temper of material.

Regulations: Compulsory rules set out to control and manage procedures within any industry.

Resistance: Deflecting the heat away and not transferring heat to a material behind or around it.

Risks: Situations that may arise, which could compromise your safety.

Siphonage: A process using a U-bend to draw fluid from one container to another at a lower level by gravity.

Sleeved: Term to describe keeping the main system pipework away from a material which may cause damage by means of using a larger diameter 'sleeve' around it.

Specification: Detailed document with an exact statement of particulars, especially a statement prescribing materials, dimensions and quality of work for something to be built, installed or manufactured.

Stack: The soil waste system pipework.

Stratification: The different layers of temperatures that form in the hot water cylinder from cold at the bottom to hot at the top.

Terminate: To come to an end of a process. For instance when running water through pipes out to a tap, the tap would be the termination point.

Thermal: Relating to or caused by heat.

Thermostat: Device used for the control of temperatures.

Threaded: Describes the jointing technique for LCS pipework.

Tight: Term sometimes used to describe a fitting that has been made correctly.

Velocity: The speed and direction of motion of a substance flowing from one point to another.

Vented: Ventilated.

Vented: When a system is not sealed and is classed as being atmospheric, working under normal conditions.

Vertical: Going straight up and down.

Volume: The amount of three-dimensional space an object occupies.

Wholesome (water): Water of good quality, fit for humans to drink.

Workplace: Where you are working. This can be a construction site, but can mean other locations, such as a customer's house, a school or a hospital.

Yorkshire (fitting): Manufacturer's name to describe a solder ring fitting.